T0176013

Foundations
of the
Theory of
Probability

Second English Edition

A. N. Kolmogorov

Translation Edited by
Nathan Morrison

With an added Bibliography by
A. T. Bharucha-Reid

Dover Publications
Garden City, New York

Bibliographical Note

This Dover edition, first published in 2018, is an unabridged republication of the 1956 second edition of the work originally published in 1950 by Chelsea Publishing Company, New York.

Library of Congress Cataloging-in-Publication Data

Names: Kolmogorov, A. N. (Andreæi Nikolaevich), 1903–1987, author. | Bharucha-Reid, A. T. (Albert T.)
Title: Foundations of the theory of probability / A.N. Kolmogorov ; translation edited by Nathan Morrison ; with an added Bibliography by A.T. Bharucha-Reid.
Other titles: Grundbegriffe der Wahrscheinlichkeitsrechnung. English
Description: Second English edition, Dover edition. | Garden City, New York : Dover Publications 2018. | Second English edition, originally published: New York : Chelsea Publishing Company, 1956; based on a title originally published: New York : Chelsea Publishing Company, 1950. | Includes bibliographical references.
Identifiers: LCCN 2017046438| ISBN 9780486821597 | ISBN 0486821595
Subjects: LCSH: Probabilities.
Classification: LCC QA273 .K614 2018 | DDC 519.2—dc23
LC record available at https://lccn.loc.gov/2017046438

Manufactured in the United States of America
82159508   2022
www.doverpublications.com

EDITOR'S NOTE

In the preparation of this English translation of Professor Kolmogorov's fundamental work, the original German monograph *Grundbegriffe der Wahrscheinlichkeitrechnung* which appeared in the Ergebnisse Der Mathematik in 1933, and also a Russian translation by G. M. Bavli published in 1936 have been used.

It is a pleasure to acknowledge the invaluable assistance of two friends and former colleagues, Mrs. Ida Rhodes and Mr. D. V. Varley, and also of my niece, Gizella Gross.

Thanks are also due to Mr. Roy Kuebler who made available for comparison purposes his independent English translation of the original German monograph.

Nathan Morrison

PREFACE

The purpose of this monograph is to give an axiomatic foundation for the theory of probability. The author set himself the task of putting in their natural place, among the general notions of modern mathematics, the basic concepts of probability theory—concepts which until recently were considered to be quite peculiar.

This task would have been a rather hopeless one before the introduction of Lebesgue's theories of measure and integration. However, after Lebesgue's publication of his investigations, the analogies between measure of a set and probability of an event, and between integral of a function and mathematical expectation of a random variable, became apparent. These analogies allowed of further extensions; thus, for example, various properties of independent random variables were seen to be in complete analogy with the corresponding properties of orthogonal functions. But if probability theory was to be based on the above analogies, it still was necessary to make the theories of measure and integration independent of the geometric elements which were in the foreground with Lebesgue. This has been done by Fréchet.

While a conception of probability theory based on the above general viewpoints has been current for some time among certain mathematicians, there was lacking a complete exposition of the whole system, free of extraneous complications. (Cf., however, the book by Fréchet, [2] in the bibliography.)

I wish to call attention to those points of the present exposition which are outside the above-mentioned range of ideas familiar to the specialist. They are the following: Probability distributions in infinite-dimensional spaces (Chapter III, § 4) ; differentiation and integration of mathematical expectations with respect to a parameter (Chapter IV, § 5) ; and especially the theory of conditional probabilities and conditional expectations (Chapter V). It should be emphasized that these new problems arose, of necessity, from some perfectly concrete physical problems.[1]

---

[1] Cf., e.g., the paper by M. Leontovich quoted in footnote 6 on p. 46; also the joint paper by the author and M. Leontovich, *Zur Statistik der kontinuierlichen Systeme und des zeitlichen Verlaufes der physikalischen Vorgänge.* Phys. Jour. of the USSR, Vol. 3, 1933, pp. 35-63.

The sixth chapter contains a survey, without proofs, of some results of A. Khinchine and the author of the limitations on the applicability of the ordinary and of the strong law of large numbers. The bibliography contains some recent works which should be of interest from the point of view of the foundations of the subject.

I wish to express my warm thanks to Mr. Khinchine, who has read carefully the whole manuscript and proposed several improvements.

Kljasma near Moscow, Easter 1933.

A. Kolmogorov

CONTENTS

V. Conditional Probabilities and Mathematical Expectations

VI. Independence; The Law of Large Numbers

Chapter I

ELEMENTARY THEORY OF PROBABILITY

We define as elementary theory of probability that part of the theory in which we have to deal with probabilities of only a finite number of events. The theorems which we derive here can be applied also to the problems connected with an infinite number of random events. However, when the latter are studied, essentially new principles are used. Therefore the only axiom of the mathematical theory of probability which deals particularly with the case of an infinite number of random events is not introduced until the beginning of Chapter II (Axiom VI).

The theory of probability, as a mathematical discipline, can and should be developed from axioms in exactly the same way as Geometry and Algebra. This means that after we have defined the elements to be studied and their basic relations, and have stated the axioms by which these relations are to be governed, all further exposition must be based exclusively on these axioms, independent of the usual concrete meaning of these elements and their relations.

In accordance with the above, in § 1 the concept of a *field of probabilities* is defined as a system of sets which satisfies certain conditions. What the elements of this set represent is of no importance in the purely mathematical development of the theory of probability (cf. the introduction of basic geometric concepts in the *Foundations of Geometry* by Hilbert, or the definitions of groups, rings and fields in abstract algebra).

Every axiomatic (abstract) theory admits, as is well known, of an unlimited number of concrete interpretations besides those from which it was derived. Thus we find applications in fields of science which have no relation to the concepts of random event and of probability in the precise meaning of these words.

The postulational basis of the theory of probability can be established by different methods in respect to the selection of axioms as well as in the selection of basic concepts and relations. However, if our aim is to achieve the utmost simplicity both in

1

the system of axioms and in the further development of the theory, then the postulational concepts of a random event and its probability seem the most suitable. There are other postulational systems of the theory of probability, particularly those in which the concept of probability is not treated as one of the basic concepts, but is itself expressed by means of other concepts.[1] However, in that case, the aim is different, namely, to tie up as closely as possible the mathematical theory with the empirical development of the theory of probability.

§ 1. Axioms[2]

Let E be a collection of elements $\xi, \eta, \zeta, \ldots$, which we shall call *elementary events*, and $\mathfrak{F}$ a set of subsets of E; the elements of the set $\mathfrak{F}$ will be called *random events*.

   I. $\mathfrak{F}$ *is a field[3] of sets.*

   II. $\mathfrak{F}$ *contains the set E.*

   III. *To each set A in $\mathfrak{F}$ is assigned a non-negative real number* $P(A)$. *This number* $P(A)$ *is called the probability of the event A.*

   IV. $P(E)$ *equals* 1.

   V. *If A and B have no element in common, then*

$$P(A+B) = P(A) + P(B)$$

A system of sets, $\mathfrak{F}$, together with a definite assignment of numbers $P(A)$, satisfying Axioms I-V, is called a *field of probability*.

Our system of Axioms I-V is *consistent*. This is proved by the following example. Let E consist of the single element ξ and let $\mathfrak{F}$ consist of E and the null set 0. $P(E)$ is then set equal to 1 and $P(0)$ equals 0.

---

[1] For example, R. von Mises[1] and [2] and S. Bernstein [1].

[2] The reader who wishes from the outset to give a concrete meaning to the following axioms, is referred to § 2.

[3] Cf. HAUSDORFF, *Mengenlehre*, 1927, p. 78. A system of sets is called a field if the sum, product, and difference of two sets of the system also belong to the same system. Every non-empty field contains the null set 0. Using Hausdorff's notation, we designate the product of A and B by AB; the sum by $A + B$ in the case where $AB = 0$; and in the general case by $A + B$; the difference of A and B by $A - B$. The set $E - A$, which is the complement of A, will be denoted by $\bar{A}$. We shall assume that the reader is familiar with the fundamental rules of operations of sets and their sums, products, and differences. All subsets of $\mathfrak{F}$ will be designated by Latin capitals.

Our system of axioms is not, however, *complete,* for in various problems in the theory of probability different fields of probability have to be examined.

The Construction of Fields of Probability. The simplest fields of probability are constructed as follows. We take an arbitrary finite set $E = \{\xi_1, \xi_2, \ldots, \xi_k\}$ and an arbitrary set $\{p_1, p_2, \ldots, p_k\}$ of non-negative numbers with the sum $p_1 + p_2 + \ldots + p_k = 1$. $\mathfrak{F}$ is taken as the set of all subsets in E, and we put

$$\mathsf{P}\{\xi_{i_1}, \xi_{i_2}, \ldots, \xi_{i_\lambda}\} = p_{i_1} + p_{i_2} + \cdots + p_{i_\lambda}.$$

In such cases, $p_1, p_2, \ldots, p_k$ are called the probabilities of the elementary events $\xi_1, \xi_2, \ldots, \xi_k$ or simply elementary probabilities. In this way are derived all possible *finite* fields of probability in which $\mathfrak{F}$ consists of the set of all subsets of E. (The field of probability is called *finite* if the set E is finite.) For further examples see Chap. II, § 3.

§ 2. The Relation to Experimental Data[4]

We apply the theory of probability to the actual world of experiments in the following manner:

1) There is assumed a complex of conditions, $\mathfrak{S}$, which allows of any number of repetitions.

2) We study a definite set of events which could take place as a result of the establishment of the conditions $\mathfrak{S}$. In individual cases where the conditions are realized, the events occur, generally, in different ways. Let E be the set of all possible variants $\xi_1, \xi_2, \ldots$ of the outcome of the given events. Some of these variants might in general not occur. We include in set E all the variants which we regard *a priori* as possible.

3) If the variant of the events which has actually occurred

---

[4] The reader who is interested in the purely mathematical development of the theory only, need not read this section, since the work following it is based only upon the axioms in § 1 and makes no use of the present discussion. Here we limit ourselves to a simple explanation of how the axioms of the theory of probability arose and disregard the deep philosophical dissertations on the concept of probability in the experimental world. In establishing the premises necessary for the applicability of the theory of probability to the world of actual events, the author has used, in large measure, the work of R. v. Mises, [1] pp. 21-27.

upon realization of conditions $\mathfrak{S}$ belongs to the set A (defined in any way), then we say that the event A has taken place.

Example: Let the complex $\mathfrak{S}$ of conditions be the tossing of a coin two times. The set of events mentioned in Paragraph 2)consists of the fact that at each toss either a head or tail may come up. From this it follows that only four different variants (elementary events) are possible, namely: *HH, HT, TH, TT*. If the "event A" connotes the occurrence of a repetition, then it will consist of a happening of either of the first or fourth of the four elementary events. In this manner, every event may be regarded as a set of elementary events.

4) Under certain conditions, which we shall not discuss here, we may assume that to an event A which may or may not occur under conditions $\mathfrak{S}$, is assigned a real number $P(A)$ which has the following characteristics:

(a) One can be practically certain that if the complex of conditions $\mathfrak{S}$ is repeated a large number of times, n, then if m be the number of occurrences of event A, the ratio m/n will differ very slightly from $P(A)$.

(b) If $P(A)$ is very small, one can be practically certain that when conditions $\mathfrak{S}$ are realized only once, the event A would not occur at all.

The Empirical Deduction of the Axioms. In general, one may assume that the system $\mathfrak{F}$ of the observed events $A, B, C, \ldots$ to which are assigned definite probabilities, form a field containing as an element the set E (Axioms I, II, and the first part of III, postulating the existence of probabilities). It is clear that $0 \leqq m/n \leqq 1$ so that the second part of Axiom III is quite natural. For the event E, m is always equal to n, so that it is natural to postulate $P(E) = 1$ (Axiom IV). If, finally, A and B are nonintersecting (incompatible), then $m = m_1 + m_2$ where m, m_1, m_2 are respectively the number of experiments in which the events $A + B, A,$ and B occur. From this it follows that

$$\frac{m}{n} = \frac{m_1}{n} + \frac{m_2}{n}.$$

It therefore seems appropriate to postulate that $P(A + B) = P(A) + P(B)$ (Axiom V).

Remark 1. If two separate statements are each practically reliable, then we may say that simultaneously they are both reliable, although the degree of reliability is somewhat lowered in the process. If, however, the number of such statements is very large, then from the practical reliability of each, one cannot deduce anything about the simultaneous correctness of all of them. Therefore from the principle stated in (a) it does not follow that in a very large number of series of n tests each, in *each* the ratio m/n will differ only slightly from P(A).

Remark 2. To an impossible event (an empty set) corresponds, in accordance with our axioms, the probability $P(0) = 0$[5], but the converse is not true: $P(A) = 0$ does not imply the impossibility of A. When $P(A) = 0$, from principle (b) all we can assert is that when the conditions $\mathfrak{S}$ are realized but once, event A is practically impossible. It does not at all assert, however, that in a sufficiently long series of tests the event A will not occur. On the other hand, one can deduce from the principle (a) merely that when $P(A) = 0$ and n is very large, the ratio m/n will be very small (it might, for example, be equal to $1/n$).

§ 3. Notes on Terminology

We have defined the objects of our future study, random events, as sets. However, in the theory of probability many set-theoretic concepts are designated by other terms. We shall give here a brief list of such concepts.

Theory of Sets	*Random Events*
1. A and B do not intersect, i.e., $AB = 0$.	1. Events A and B are incompatible.
2. $AB...N = 0$.	2. Events $A, B, ..., N$ are incompatible.
3. $AB...N = X$.	3. Event X is defined as the simultaneous occurrence of events $A, B, ..., N$.
4. $A \overset{.}{+} B \overset{.}{+} ... \overset{.}{+} N = X$.	4. Event X is defined as the occurrence of at least one of the events $A, B, ..., N$.

---

[5] Cf. § 4, Formula (3).

Theory of Sets	*Random Events*
5. The complementary set $\bar{A}$.	5. The opposite event $\bar{A}$ consisting of the non-occurrence of event A.
6. $A = 0$.	6. Event A is impossible.
7. $A = E$.	7. Event A must occur.
8. The system $\mathfrak{A}$ of the sets $A_1, A_2, \ldots, A_n$ forms a *decomposition* of the set E if $A_1 + A_2 + \ldots + A_n = E$. (This assumes that the sets A_i do not intersect, in pairs.)	8. *Experiment* $\mathfrak{A}$ consists of determining which of the events $A_1, A_2, \ldots, A_n$ occurs. We therefore call $A_1, A_2, \ldots, A_n$ the possible results of experiment $\mathfrak{A}$.
9. B is a subset of $A: B \subset A$.	9. From the occurrence of event B follows the inevitable occurrence of A.

§ 4. Immediate Corollaries of the Axioms; Conditional Probabilities; Theorem of Bayes

From $A + \bar{A} = E$ and the Axioms IV and V it follows that

$$P(A) + P(\bar{A}) = 1 \qquad (1)$$

$$P(\bar{A}) = 1 - P(A) . \qquad (2)$$

Since $\bar{E} = 0$, then, in particular,

$$P(0) = 0 . \qquad (3)$$

If $A, B, \ldots, N$ are incompatible, then from Axiom V follows the formula (the Addition Theorem)

$$P(A + B + \ldots + N) = P(A) + P(B) + \ldots + P(N) . \qquad (4)$$

If $P(A) > 0$, then the quotient

$$P_A(B) = \frac{P(AB)}{P(A)} \qquad (5)$$

is defined to be the *conditional probability* of the event B under the condition A.

From (5) it follows immediately that

$$P(AB) = P(A)\,\mathsf{P}_A(B). \tag{6}$$

And by induction we obtain the general formula (the Multiplication Theorem)

$$P(A_1 A_2 \ldots A_n) = P(A_1)\,\mathsf{P}_{A_1}(A_2)\,\mathsf{P}_{A_1 A_2}(A_3)\ldots \mathsf{P}_{A_1 A_2 \ldots A_{n-1}}(A_n). \tag{7}$$

The following theorems follow easily:

$$\mathsf{P}_A(B) \geqq 0, \tag{8}$$

$$\mathsf{P}_A(E) = 1, \tag{9}$$

$$\mathsf{P}_A(B + C) = \mathsf{P}_A(B) + \mathsf{P}_A(C). \tag{10}$$

Comparing formulae (8)—(10) with axioms III—V, we find that the system $\mathfrak{F}$ of sets together with the set function $\mathsf{P}_A(B)$ (provided A is a fixed set), form a field of probability and therefore, *all the above general theorems concerning* $P(B)$ *hold true for the conditional probability* $\mathsf{P}_A(B)$ (provided the event A is fixed).

It is also easy to see that

$$\mathsf{P}_A(A) = 1. \tag{11}$$

From (6) and the analogous formula

$$P\ (AB) = P(B)\,\mathsf{P}_B(A)$$

we obtain the important formula:

$$\mathsf{P}_B(A) = \frac{P(A)\,\mathsf{P}_A(B)}{P(B)}, \tag{12}$$

which contains, in essence, the Theorem of Bayes.

THE THEOREM ON TOTAL PROBABILITY: Let $A_1 + A_2 + \ldots + A_n = E$ (this assumes that the events $A_1, A_2, \ldots, A_n$ are mutually exclusive) and let X be arbitrary. Then

$$P(X) = P(A_1)\,\mathsf{P}_{A_1}(X) + P(A_2)\,\mathsf{P}_{A_2}(X) + \cdots + P(A_n)\,\mathsf{P}_{A_n}(X). \tag{13}$$

Proof:

$$X = A_1 X + A_2 X + \ldots + A_n X;$$

using (4) we have

$$P(X) = P(A_1 X) + P(A_2 X) + \ldots + P(A_n X)$$

and according to (6) we have at the same time

$$P(A_i X) = P(A_i)\,\mathsf{P}_{A_i}(X).$$

THE THEOREM OF BAYES: Let $A_1 + A_2 + \ldots + A_n = E$ and X be arbitrary, then

$$\left.\mathsf{P}_X(A_i) = \frac{P(A_i)\,\mathsf{P}_{A_i}(X)}{P(A_1)\,\mathsf{P}_{A_1}(X) + P(A_2)\,\mathsf{P}_{A_2}(X) + \cdots + P(A_n)\,\mathsf{P}_{A_n}(X)},\right\} \tag{14}$$
$$i = 1, 2, 3, \ldots, n.$$

A_1, A_2, ... , A_n are often called "hypotheses" and formula
(14) is considered as the probability $P_X(A_i)$ of the hypothesis
A_i after the occurrence of event X. [$P(A_i)$ then denotes the
a priori probability of A_i.]

Proof: From (12) we have

$$P_X(A_i) = \frac{P(A_i)\,P_{A_i}(X)}{P(X)}.$$

To obtain the formula (14) it only remains to substitute for the
probability $P(X)$ its value derived from (13) by applying the
theorem on total probability.

§ 5. Independence

The concept of mutual *independence* of two or more experiments holds, in a certain sense, a central position in the theory of
probability. Indeed, as we have already seen, the theory of
probability can be regarded from the mathematical point of view
as a special application of the general theory of additive set functions. One naturally asks, how did it happen that the theory of
probability developed into a large individual science possessing
its own methods?

In order to answer this question, we must point out the spe-
cialization undergone by general problems in the theory of additive set functions when they are proposed in the theory of
probability.

The fact that our additive set function $P(A)$ is non-negative
and satisfies the condition $P(E) = 1$, does not in itself cause new
difficulties. Random variables (see Chap. III) from a mathematical point of view represent merely functions measurable with
respect to $P(A)$, while their mathematical expectations are
abstract Lebesgue integrals. (This analogy was explained fully
for the first time in the work of Fréchet[6].) The mere introduction
of the above concepts, therefore, would not be sufficient to pro-
duce a basis for the development of a large new theory.

Historically, the independence of experiments and random
variables represents the very mathematical concept that has given
the theory of probability its peculiar stamp. The classical work
or LaPlace, Poisson, Tchebychev, Markov, Liapounov, Mises, and

---

[6] See Fréchet [1] and [2].

Bernstein is actually dedicated to the fundamental investigation of series of independent random variables. Though the latest dissertations (Markov, Bernstein and others) frequently fail to assume complete independence, they nevertheless reveal the necessity of introducing analogous, weaker, conditions, in order to obtain sufficiently significant results (see in this chapter § 6, Markov chains).

We thus see, in the concept of independence, at least the germ of the peculiar type of problem in probability theory. In this book, however, we shall not stress that fact, for here we are interested mainly in the logical foundation for the specialized investigations of the theory of probability.

In consequence, one of the most important problems in the philosophy of the natural sciences is—in addition to the well-known one regarding the essence of the concept of probability itself—to make precise the premises which would make it possible to regard any given real events as independent. This question, however, is beyond the scope of this book.

Let us turn to the definition of independence. Given n experiments $\mathfrak{A}^{(1)}$, $\mathfrak{A}^{(2)}$, . . . , $\mathfrak{A}^{(n)}$, that is, n decompositions

$$E = A_1^{(i)} + A_2^{(i)} + \cdots + A_{r_i}^{(i)} \qquad i = 1, 2, \ldots, n$$

of the basic set E. It is then possible to assign $r = r_1 r_2 \ldots r_n$ probabilities (in the general case)

$$p_{q_1 q_2 \ldots q_n} = \mathsf{P}\left(A_{q_1}^{(1)} A_{q_2}^{(2)} \ldots A_{q_n}^{(n)}\right) \geqq 0$$

which are entirely arbitrary except for the single condition[7] that

$$\sum_{q_1, q_2, \ldots, q_n} p_{q_1 q_2 \ldots q_n} = 1 \tag{1}$$

DEFINITION I. n experiments $\mathfrak{A}^{(1)}$, $\mathfrak{A}^{(2)}$, . . . , $\mathfrak{A}^{(n)}$ are called *mutually independent*, if for any q_1, q_2, . . . , q_n the following equation holds true:

$$\mathsf{P}\left(A_{q_1}^{(1)} A_{q_2}^{(2)} \ldots A_{q_n}^{(n)}\right) = \mathsf{P}\left(A_{q_1}^{(1)}\right) \mathsf{P}\left(A_{q_2}^{(2)}\right) \ldots \mathsf{P}\left(A_{q_n}^{(n)}\right) . \tag{2}$$

---

[7] One may construct a field of probability with arbitrary probabilities subject only to the above-mentioned conditions, as follows: E is composed of r elements $\xi_{q_1 q_2 \ldots q_n}$. Let the corresponding elementary probabilities be $p_{q_1 q_2 \ldots q_n}$, and finally let $A_q^{(i)}$ be the set of all $\xi_{q_1 q_2 \ldots q_n}$ for which $q_i = q$.

Among the r equations in (2), there are only $r-r_1-r_2-\ldots-r_n+n-1$ independent equations[8].

THEOREM I. If n experiments $\mathfrak{A}^{(1)}$, $\mathfrak{A}^{(2)}$, $\ldots$, $\mathfrak{A}^{(n)}$ are mutually independent, then any m of them $(m < n)$, $\mathfrak{A}^{(i_1)}$, $\mathfrak{A}^{(i_2)}$, $\ldots$, $\mathfrak{A}^{(i_m)}$, are also independent[9].

In the case of independence we then have the equations:

$$P\left(A_{q_1}^{(i_1)} A_{q_2}^{(i_2)} \ldots A_{q_m}^{(i_m)}\right) = P\left(A_{q_1}^{(i_1)}\right) P\left(A_{q_2}^{(i_2)}\right) \ldots P\left(A_{q_m}^{(i_m)}\right) \qquad (3)$$

(all i_k must be different.)

DEFINITION II. n events A_1, A_2, $\ldots$, A_n are *mutually independent*, if the decompositions (trials)

$$E = A_k + \bar{A}_k \qquad (k = 1, 2, \ldots, n)$$

are independent.

In this case $r_1 = r_2 = \ldots = r_n = 2, r = 2^n$; therefore, of the 2^n equations in (2) only $2^n - n - 1$ are independent. The necessary and sufficient conditions for the independence of the events A_1, A_2, $\ldots$, A_n are the following $2^n - n - 1$ equations[10]:

$$P(A_{i_1} A_{i_2} \ldots A_{i_m}) = P(A_{i_1}) P(A_{i_2}) \ldots P(A_{i_m}), \qquad (4)$$
$$m = 1, 2, \ldots, n,$$
$$1 \leqq i_1 < i_2 < \cdots < i_m \leqq n.$$

All of these equations are mutually independent.

In the case $n = 2$ we obtain from (4) only one condition $(2^2 - 2 -$

---

[8] Actually, in the case of independence, one may choose arbitrarily only $r_1 + r_2 + \ldots + r_n$ probabilities $p_q^{(i)} = P(A_q^{(i)})$ so as to comply with the n conditions

$$\sum_q p_q^{(i)} = 1.$$

Therefore, in the general case, we have $r-1$ degrees of freedom, but in the case of independence only $r_1 + r_2 + \ldots + r_n - n$.

[9] To prove this it is sufficient to show that from the mutual independence of n decompositions follows the mutual independence of the first $n-1$. Let us assume that the equations (2) hold. Then

$$P\left(A_{q_1}^{(1)} A_{q_2}^{(2)} \ldots A_{q_{n-1}}^{(n-1)}\right) = \sum_{q_n} P\left(A_{q_1}^{(1)} A_{q_2}^{(2)} \ldots A_{q_n}^{(n)}\right)$$

$$= P\left(A_{q_1}^{(1)}\right) P\left(A_{q_2}^{(2)}\right) \ldots P\left(A_{q_{n-1}}^{(n-1)}\right) \sum_{q_n} P\left(A_{q_n}^{(n)}\right) = P\left(A_{q_1}^{(1)}\right) P\left(A_{q_2}^{(2)}\right) \ldots P\left(A_{q_{n-1}}^{(n-1)}\right),$$

Q.E.D.

[10] See S. N. Bernstein [1] pp. 47-57. However, the reader can easily prove this himself (using mathematical induction).

1 = 1) for the independence of two events A_1 and A_2:

$$P(A_1 A_2) = P(A_1) P(A_2). \tag{5}$$

The system of equations (2) reduces itself, in this case, to three equations, besides (5) :

$$P(A_1 \bar{A}_2) = P(A_1) P(\bar{A}_2)$$
$$P(\bar{A}_1 A_2) = P(\bar{A}_1) P(A_2)$$
$$P(\bar{A}_1 \bar{A}_2) = P(\bar{A}_1) P(\bar{A}_2) \quad,$$

which obviously follow from (5).[11]

It need hardly be remarked that from the independence of the events $A_1, A_2, \ldots, A_n$ *in pairs*, i.e. from the relations

$$P(A_i A_j) = P(A_i) P(A_j) \qquad (i \neq j)$$

it does not at all follow that when $n > 2$ these events are independent[12]. (For that we need the existence of all equations (4).)

In introducing the concept of independence, no use was made of conditional probability. Our aim has been to explain as clearly as possible, in a purely mathematical manner, the meaning of this concept. Its applications, however, generally depend upon the properties of certain conditional probabilities.

If we assume that all probabilities $P(A_q^{(i)})$ are positive, then from the equations (3) it follows[13] that

$$P_{A_{q_1}^{(i_1)} A_{q_2}^{(i_2)} \ldots A_{q_{m-1}}^{(i_{m-1})}} \left(A_{q_m}^{(i_m)} \right) = P \left(A_{q_m}^{(i_m)} \right). \tag{6}$$

From the fact that formulas (6) hold, and from the Multiplication Theorem (Formula (7), § 4), follow the formulas (2). We obtain, therefore,

THEOREM II: *A necessary and sufficient condition for independence of experiments* $\mathfrak{A}^{(1)}, \mathfrak{A}^{(2)}, \ldots, \mathfrak{A}^{(n)}$ *in the case of posi-*

---

[11] $P(A_1 \bar{A}_2) = P(A_1) - P(A_1 A_2) = P(A_1) - P(A_1) P(A_2) = P(A_1) \{1 - P(A_2)\}$ $= P(A_1) P(\bar{A}_2)$, etc.

[12] This can be shown by the following simple example (S. N. Bernstein) : Let set E be composed of four elements $\xi_1, \xi_2, \xi_3, \xi_4$; the corresponding elementary probabilities p_1, p_2, p_3, p_4 are each assumed to be ¼ and

$$A = \{\xi_1, \xi_2\}, \quad B = \{\xi_1, \xi_3\}, \quad C = \{\xi_1, \xi_4\}.$$

It is easy to compute that
$$P(A) = P(B) = P(C) = \tfrac{1}{2},$$
$$P(AB) = P(BC) = P(AC) = \tfrac{1}{4} = (\tfrac{1}{2})^2,$$
$$P(ABC) = \tfrac{1}{4} \neq (\tfrac{1}{2})^3.$$

[13] To prove it, one must keep in mind the definition of conditional probability (Formula (5), § 4) and substitute for the probabilities of products the products of probabilities according to formula (3).

tive probabilities $\mathsf{P}(A_q^{(i)})$ *is that the conditional probability of the results* $A_q{}^{(i)}$ *of experiments* $\mathfrak{A}^{(i)}$ *under the hypothesis that several other tests* $\mathfrak{A}^{(i_1)}, \mathfrak{A}^{(i_2)}, \dots, \mathfrak{A}^{(i_k)}$ *have had definite results* $A_{q_1}^{(i_1)}, A_{q_2}^{(i_2)}, A_{q_3}^{(i_3)}, \dots, A_{q_k}^{(i_k)}$ *is equal to the absolute probability* $\mathsf{P}(A_q^{(i)})$.

On the basis of formulas (4) we can prove in an analogous manner the following theorem:

THEOREM III. *If all probabilities* $\mathsf{P}(A_k)$ *are positive, then a necessary and sufficient condition for mutual independence of the events* $A_1, A_2, \dots, A_n$ *is the satisfaction of the equations*

$$\mathsf{P}_{A_{i_1} A_{i_2} \cdots A_{i_k}}(A_i) = \mathsf{P}(A_i) \qquad (7)$$

for any pairwise different $i_1, i_2, \dots, i_k, i$.

In the case $n = 2$ the conditions (7) reduce to two equations:

$$\left.\begin{array}{l} \mathsf{P}_{A_1}(A_2) = \mathsf{P}(A_2), \\ \mathsf{P}_{A_2}(A_1) = \mathsf{P}(A_1). \end{array}\right\} \qquad (8)$$

It is easy to see that the first equation in (8) alone is a necessary and sufficient condition for the independence of A_1 and A_2 provided $\mathsf{P}(A_1) > 0$.

§ 6. Conditional Probabilities as Random Variables, Markov Chains

Let $\mathfrak{A}$ be a decomposition of the fundamental set E:

$$E = A_1 + A_2 + \dots + A_r,$$

and x a real function of the elementary event ξ, which for every set A_q is equal to a corresponding constant a_q. x is then called a *random variable*, and the sum

$$E(x) = \sum_q a_q \mathsf{P}(A_q)$$

is called the *mathematical expectation* of the variable x. The theory of random variables will be developed in Chaps. III and IV. We shall not limit ourselves there merely to those random variables which can assume only a finite number of different values.

A random variable which for every set A_q assumes the value $\mathsf{P}_{A_{q_i}}(B)$, we shall call *the conditional probability of the event B after the given experiment $\mathfrak{A}$* and shall designate it by $\mathsf{P}_{\mathfrak{A}}(B)$. Two experiments $\mathfrak{A}^{(1)}$ and $\mathfrak{A}^{(2)}$ are independent if, and only if,

$$P_{\mathfrak{A}^{(1)}}(A_q^{(2)}) = P(A_q^{(2)}) \qquad q = 1, 2, \ldots, r_2.$$

Given any decompositions (experiments) $\mathfrak{A}^{(1)}, \mathfrak{A}^{(2)}, \ldots, \mathfrak{A}^{(n)}$, we we shall represent by

$$\mathfrak{A}^{(1)}\mathfrak{A}^{(2)} \ldots \mathfrak{A}^{(n)}$$

the decomposition of set E into the products

$$A_{q_1}^{(1)} A_{q_2}^{(2)} \ldots A_{q_n}^{(n)}.$$

Experiments $\mathfrak{A}^{(1)}, \mathfrak{A}^{(2)}, \ldots, \mathfrak{A}^{(n)}$ are mutually independent when and only when

$$P_{\mathfrak{A}^{(1)} \mathfrak{A}^{(2)} \ldots \mathfrak{A}^{(k-1)}}(A_q^{(k)}) = P(A_q^{(k)}),$$

k and q being arbitrary[14].

DEFINITION: The sequence $\mathfrak{A}^{(1)}, \mathfrak{A}^{(2)}, \ldots, \mathfrak{A}^{(n)}, \ldots$ forms a Markov chain if for arbitrary n and q

$$P_{\mathfrak{A}^{(1)} \mathfrak{A}^{(2)} \ldots \mathfrak{A}^{(n-1)}}(A_q^{(n)}) = P_{\mathfrak{A}^{(n-1)}}(A_q^{(n)}).$$

Thus, Markov chains form a natural generalization of sequences of mutually independent experiments. If we set

$$p_{q_m q_n}(m, n) = P_{A_{q_m}^{(m)}}(A_{q_n}^{(n)}) \qquad m < n ,$$

then the basic formula of the theory of Markov chains will assume the form:

$$p_{q_k q_n}(k, n) = \sum_{q_m} p_{q_k q_m}(k, m)\, p_{q_m q_n}(m, n), \qquad k < m < n. \tag{1}$$

If we denote the matrix $\|p_{q_m q_n}(m, n)\|$ by $p(m, n)$, (1) can be written as[15]:

$$p(k,n) = p(k,m)\, p(m,n) \qquad k < m < n. \tag{2}$$

---

[14] The necessity of these conditions follows from Theorem II, § 5; that they are also sufficient follows immediately from the Multiplication Theorem (Formula (7) of § 4).

[15] For further development of the theory of Markov chains, see R. v. Mises [1], § 16, and B. HOSTINSKY, *Méthodes générales du calcul des probabilités*, "Mém. Sci. Math." V. 52, Paris 1931.

Chapter II

INFINITE PROBABILITY FIELDS

§ 1. Axiom of Continuity

We denote by $\underset{m}{\mathfrak{D}} A_m$, as is customary, the product of the sets A_m (whether finite or infinite in number) and their sum by $\underset{m}{\mathfrak{S}} A_m$. Only in the case of disjoint sets A_m is the form $\sum_m A_m$ used instead of $\underset{m}{\mathfrak{S}} A_m$. Consequently,

$$\underset{m}{\mathfrak{S}} A_m = A_1 \dotplus A_2 \dotplus \cdots,$$

$$\sum_m A_m = A_1 + A_2 + \cdots,$$

$$\underset{m}{\mathfrak{D}} A_m = A_1 A_2 \cdots.$$

In all future investigations, we shall assume that besides Axioms I - V, still another holds true:

VI. *For a decreasing sequence of events*

$$A_1 \supset A_2 \supset \cdots \supset A_n \supset \cdots \tag{1}$$

of $\mathfrak{F}$, for which

$$\underset{n}{\mathfrak{D}} A_n = 0 \ , \tag{2}$$

the following equation holds:

$$\lim \mathsf{P}(A_n) = 0. \qquad\qquad n \to \infty \tag{3}$$

In the future we shall designate by *probability field* only a field of probability as outlined in the first chapter, which also satisfies Axiom VI. The fields of probability as defined in the first chapter without Axiom VI might be called *generalized fields of probability*.

If the system $\mathfrak{F}$ of sets is finite, Axiom VI follows from Axioms I - V. For actually, in that case there exist only a finite number of different sets in the sequence (1). Let A_k be the smallest among them, then all sets A_{k+p} coincide with A_k and we obtain then

$$A_k = A_{k+p} = \mathop{\mathfrak{D}}_{n} A_n = 0,$$

$$\lim \mathsf{P}(A_n) = \mathsf{P}(0) = 0.$$

All examples of *finite* fields of probability, in the first chapter, satisfy, therefore, Axiom VI. The system of Axioms I - VI then proves to be *consistent* and *incomplete*.

For infinite fields, on the other hand, the Axiom of Continuity, VI, proved to be independent of Axioms I - V. Since the new axiom is essential for infinite fields of probability only, it is almost impossible to elucidate its empirical meaning, as has been done, for example, in the case of Axioms I - V in § 2 of the first chapter. For, in describing any observable random process we can obtain only finite fields of probability. Infinite fields of probability occur only as idealized models of real random processes. *We limit ourselves, arbitrarily, to only those models which satisfy Axiom VI.* This limitation has been found expedient in researches of the most diverse sort.

GENERALIZED ADDITION THEOREM : *If A_1, A_2, . . . , A_n, . . . and A belong to $\mathfrak{F}$, then from*

$$A = \sum_{n} A_n \tag{4}$$

follows the equation

$$\mathsf{P}(A) = \sum_{n} \mathsf{P}(A_n). \tag{5}$$

Proof : Let

$$R_n = \sum_{m > n} A_m.$$

Then, obviously $\mathop{\mathfrak{D}}_{n}(R_n) = 0,$

and, therefore, according to Axiom VI

$$\lim \mathsf{P}(R_n) = 0 \qquad\qquad n \to \infty. \tag{6}$$

On the other hand, by the addition theorem

$$\mathsf{P}(A) = \mathsf{P}(A_1) + \mathsf{P}(A_2) + \ldots + \mathsf{P}(A_n) + \mathsf{P}(R_n). \tag{7}$$

From (6) and (7) we immediately obtain (5).

We have shown, then, *that the probability $\mathsf{P}(A)$ is a completely additive set function on $\mathfrak{F}$.* Conversely, Axioms V and VI hold true for every completely additive set function defined on any field $\mathfrak{F}$.* We can, therefore, define the concept of a field of probability in the following way: *Let E be an arbitrary set, $\mathfrak{F}$ a field of subsets of E, containing E, and P(A) a non-negative completely additive set function defined on $\mathfrak{F}$; the field $\mathfrak{F}$ together with the set function P(A) forms a field of probability.*

A COVERING THEOREM: *If $A, A_1, A_2, \ldots, A_n, \ldots$ belong to $\mathfrak{F}$ and*

$$A \subset \mathop{\mathfrak{S}}_{n} A_n \tag{8}$$

then

$$P(A) \leqq \sum_n P(A_n). \tag{9}$$

Proof:

$$A = A \mathop{\mathfrak{S}}_{n}(A_n) = A A_1 + A(A_2 - A_2 A_1) + A(A_3 - A_3 A_2 - A_3 A_1) + \cdots,$$

$$P(A) = P(A A_1) + P\{A(A_2 - A_2 A_1)\} + \cdots \leqq P(A_1) + P(A_2) + \cdots.$$

§ 2. Borel Fields of Probability

The field $\mathfrak{F}$ is called a *Borel field*, if all countable sums $\sum A_n$ of the sets A_n from $\mathfrak{F}$ belong to $\mathfrak{F}$. Borel fields are also called completely additive systems of sets. From the formula

$$\mathop{\mathfrak{S}}_{n} A_n = A_1 + (A_2 - A_2 A_1) + (A_3 - A_3 A_2 - A_3 A_1) + \cdots \tag{1}$$

we can deduce that a Borel field contains also all the sums $\mathop{\mathfrak{S}}_{n} A_n$ composed of a countable number of sets A_n belonging to it. From the formula

$$\mathop{\mathfrak{D}}_{n} A_n = E - \mathop{\mathfrak{S}}_{n} \bar{A}_n \tag{2}$$

the same can be said for the product of sets.

A field of probability is a Borel field of probability if the corresponding field $\mathfrak{F}$ is a Borel field. Only in the case of Borel fields of probability do we obtain full freedom of action, without danger of the occurrence of events having no probability. We shall now prove that we may limit ourselves to the investigation of Borel fields of probability. This will follow from the so-called extension theorem, to which we shall now turn.

Given a field of probability $(\mathfrak{F}, P)$. As is known[1], there exists a smallest Borel field $B\mathfrak{F}$ containing $\mathfrak{F}$. And we have the

---

* See, for example, O. NIKODYM, *Sur une généralisation des intégrales de M. J. Radon*, Fund. Math. v. 15, 1930, p. 136.

[1] HAUSDORFF, *Mengenlehre*, 1927, p. 85.

EXTENSION THEOREM: *It is always possible to extend a non-negative completely additive set function* P(A), *defined in* $\mathfrak{F}$, *to all sets of* $B\mathfrak{F}$ *without losing either of its properties (non-negativeness and complete additivity) and this can be done in only one way.*

The extended field $B\mathfrak{F}$ forms with the extended set function P(A) a field of probability ($B\mathfrak{F}$, P). This field of probability ($B\mathfrak{F}$, P) we shall call the *Borel extension of the field* ($\mathfrak{F}$, P)

The proof of this theorem, which belongs to the theory of additive set functions and which sometimes appears in other forms, can be given as follows:

Let A be any subset of E; we shall denote by P*(A) the lower limit of the sums

$$\sum_n P(A_n)$$

for all coverings

$$A \subset \mathop{\mathfrak{S}}_n A_n$$

of the set A by a finite or countable number of sets A_n of $\mathfrak{F}$. It is easy to prove that P*(A) is then an outer measure in the Carathéodory sense[2]. In accordance with the Covering Theorem (§ 1), P*(A) coincides with P(A) for all sets of $\mathfrak{F}$. It can be further shown that all sets of $\mathfrak{F}$ are measurable in the Carathéodory sense. Since all measurable sets form a Borel field, all sets of $B\mathfrak{F}$ are consequently measurable. The set function P*(A) is, therefore, completely additive on $B\mathfrak{F}$, and on $B\mathfrak{F}$ we may set

$$P(A) = P*(A).$$

We have thus shown the existence of the extension. The uniqueness of this extension follows immediately from the minimal property of the field $B\mathfrak{F}$.

Remark: Even if the sets (events) A of $\mathfrak{F}$ can be interpreted as actual and (perhaps only approximately) observable events, it does not, of course, follow from this that the sets of the extended field $B\mathfrak{F}$ reasonably admit of such an interpretation.

Thus there is the possibility that while a field of probability ($\mathfrak{F}$, P) may be regarded as the image (idealized, however) of

---

[2] CARATHÉODORY, *Vorlesungen über reelle Funktionen*, pp. 237-258. (New York, *Chelsea Publishing Company*).

actual random events, the extended field of probability $(B\mathfrak{F},\ \mathsf{P})$ will still remain merely a mathematical structure.

Thus sets of $B\mathfrak{F}$ are generally merely ideal events to which nothing corresponds in the outside world. However, if reasoning which utilizes the probabilities of such ideal events leads us to a determination of the probability of an actual event of $\mathfrak{F}$, then, from an empirical point of view also, this determination will automatically fail to be contradictory.

§ 3. Examples of Infinite Fields of Probability

I. In § 1 of the first chapter, we have constructed various finite probability fields.

Let now $E = \{\xi_1,\ \xi_2,\ \ldots,\ \xi_n,\ \ldots\}$ be a countable set, and let $\mathfrak{F}$ coincide with the aggregate of the subsets of E.

All possible probability fields with such an aggregate $\mathfrak{F}$ are obtained in the following manner:

We take a sequence of non-negative numbers p_n, such that

$$p_1 + p_2 + \ldots + p_n + \ldots = 1$$

and for each set A put

$$\mathsf{P}(A) = \sum_n{}' p_n,$$

where the summation $\sum'$ extends to all the indices n for which ξ_n belongs to A. These fields of probability are obviously Borel fields.

II. In this example, we shall assume that E represents the real number axis. At first, let $\mathfrak{F}$ be formed of all possible finite sums of half-open intervals $[a;\ b) = \{a \leqq \xi < b\}$ (taking into consideration not only the proper intervals, with finite a and b, but also the improper intervals $[-\infty;\ a)$, $[a;\ +\infty)$ and $[-\infty;\ +\infty)$). $\mathfrak{F}$ is then a field. By means of the extension theorem, however, each field of probability on $\mathfrak{F}$ can be extended to a similar field on $B\mathfrak{F}$. The system of sets $B\mathfrak{F}$ is, therefore, in our case nothing but the system of all Borel point sets on a line. Let us turn now to the following case.

III. Again suppose E to be the real number axis, while $\mathfrak{F}$ is composed of all Borel point sets of this line. In order to construct a field of probability with the given field $\mathfrak{F}$, it is sufficient to define an arbitrary non-negative completely additive set-function

$P(A)$ on $\mathfrak{F}$ which satisfies the condition $P(E) = 1$. As is well known,[3] such a function is uniquely determined by its values

$$P[-\infty; x) = F(x) \tag{1}$$

for the special intervals $[-\infty; x)$. The function $F(x)$ is called the *distribution function of* ξ. Further on (Chap. III, § 2) we shall shown that $F(x)$ is non-decreasing, continuous on the left, and has the following limiting values:

$$\lim_{x \to -\infty} F(x) = F(-\infty) = 0, \qquad \lim_{x \to +\infty} F(x) = F(+\infty) = 1. \tag{2}$$

Conversely, if a given function $F(x)$ satisfies these conditions, then it always determines a non-negative completely additive set-function $P(A)$ for which $P(E) = 1$.[4]

IV. Let us now consider the basic set E as an n-dimensional Euclidian space R^n, i.e., the set of all ordered n-tuples $\xi = \{x_1, x_2, \ldots, x_n\}$ of real numbers. Let $\mathfrak{F}$ consist, in this case, of all Borel point-sets[5] of the space R^n. On the basis of reasoning analogous to that used in Example II, we need not investigate narrower systems of sets, for example the systems of n-dimensional intervals.

The role of probability function $P(A)$ will be played here, as always, by any non-negative and completely additive set-function defined on $\mathfrak{F}$ and satisfying the condition $P(E) = 1$. Such a set-function is determined uniquely if we assign its values

$$P(L_{a_1 a_2 \ldots a_n}) = F(a_1, a_2, \ldots, a_n) \tag{3}$$

for the special sets $L_{a_1 a_2 \ldots a_n}$, where $L_{a_1 a_2 \ldots a_n}$ represents the aggregate of all ξ for which $x_i < a_i$ $(i = 1, 2, \ldots, n)$.

For our function $F(a_1, a_2, \ldots, a_n)$ we may choose any function which for each variable is non-decreasing and continuous on the left, and which satisfies the following conditions:

$$\left. \begin{array}{l} \displaystyle\lim_{a_i \to -\infty} F(a_1, a_2, \ldots, a_n) = F(a_1, \ldots, a_{i-1}, -\infty, a_{i+1}, \ldots, a_n) = 0, \\ \hspace{8cm} i = 1, 2, \ldots, n \\ \displaystyle\lim F(a_1, a_2, \ldots, a_n) = F(+\infty, +\infty, \ldots, +\infty) = 1. \end{array} \right\} \tag{4}$$

$$\sum_{\substack{i=1 \\ \varepsilon_i = 0,1}}^{n} (-1)^{\varepsilon_1 + \varepsilon_2 + \cdots + \varepsilon_n} F(a_1 - \varepsilon_1 c_1, a_2 - \varepsilon_2 c_2, \ldots, a_n - \varepsilon_n c_n) \geqq 0,$$

$$c_i > 0, \quad i = 1, 2, 3, \ldots, n.$$

---

[3] Cf., for example, LEBESGUE, *Leçons sur l'intégration*, 1928, p. 152-156.

[4] See the previous note.

[5] For a definition of Borel sets in R see HAUSDORFF, *Mengenlehre*, 1927, pp. 177-181.

$F(a_1, a_2, \ldots, a_n)$ is called the *distribution function* of the variables $x_1, x_2, \ldots, x_n$.

The investigation of fields of probability of the above type is sufficient for all classical problems in the theory of probability[6]. In particular, a probability function in R^n can be defined thus:

We take any non-negative point function $f(x_1, x_2, \ldots, x_n)$ defined in R^n, such that

$$\int\limits_{-\infty}^{+\infty} \int\limits_{-\infty}^{+\infty} \cdots \int\limits_{-\infty}^{+\infty} f(x_1, x_2, \ldots, x_n)\, dx_1\, dx_2 \ldots dx_n = 1$$

and set

$$\mathsf{P}(A) = \int\!\!\int \cdots \int\limits_{A} f(x_1, x_2, \ldots, x_n)\, dx_1\, dx_2 \ldots dx_n \ . \qquad (5)$$

$f(x_1, x_2, \ldots, x_n)$ is, in this case, the *probability density* at the point $(x_1, x_2, \ldots, x_n)$ (cf. Chap. III, § 2).

Another type of probability function in R^n is obtained in the following manner: Let $\{\xi_i\}$ be a sequence of points of R^n, and let $\{p_i\}$ be a sequence of non-negative real numbers, such that $\sum p_i = 1$; we then set, as we did in Example I,

$$\mathsf{P}(A) = \sum{}' p_i,$$

where the summation $\sum'$ extends over all indices i for which ξ belongs to A. The two types of probability functions in R^n mentioned here do not exhaust all possibilities, but are usually considered sufficient for applications of the theory of probability. Nevertheless, we can imagine problems of interest for applications outside of this classical region in which elementary events are defined by means of an infinite number of coordinates. The corresponding fields of probability we shall study more closely after introducing several concepts needed for this purpose. (Cf. Chap. III, § 3).

---

[6] Cf., for example, R. v. Mises [1], pp. 13-19. Here the existence of probabilities for "all practically possible" sets of an n-dimensional space is required.

Chapter III

RANDOM VARIABLES

§ 1. Probability Functions

Given a mapping of the set E into a set E' consisting of any type of elements, i.e., a single-valued function $u(\xi)$ defined on E, whose values belong to E'. To each subset A' of E' we shall put into correspondence, as its *pre-image* in E, the set $u^{-1}(A')$ of all elements of E which map onto elements of A'. Let $\mathfrak{F}^{(u)}$ be the system of all subsets A' of E', whose pre-images belong to the field $\mathfrak{F}$. $\mathfrak{F}^{(u)}$ will then also be a field. If $\mathfrak{F}$ happens to be a Borel field, the same will be true of $\mathfrak{F}^{(u)}$. We now set

$$\mathsf{P}^{(u)}(A') = \mathsf{P}\ \{u^{-1}(A')\}. \tag{1}$$

Since this set-function $\mathsf{P}^{(u)}$, defined on $\mathfrak{F}^{(u)}$, satisfies with respect to the field $\mathfrak{F}^{(u)}$ all of our Axioms I - VI, it represents a probability function on $\mathfrak{F}^{(u)}$. Before turning to the proof of all the facts just stated, we shall formulate the following definition.

DEFINITION. Given a single-valued function $u(\xi)$ of a random event ξ. The function $\mathsf{P}^{(u)}(A')$, defined by (1), is then called the *probability function* of u.

Remark 1: In studying fields of probability $(\mathfrak{F}, \mathsf{P})$, we call the function $\mathsf{P}(A)$ simply the probability function, but $\mathsf{P}^{(u)}(A')$ is called the probability function of u. In the case $u(\xi) = \xi$, $\mathsf{P}^{(u)}(A')$ coincides with $\mathsf{P}(A)$.

Remark 2: The event $u^{-1}(A')$ consists of the fact that $u(\xi)$ belongs to A'. Therefore, $\mathsf{P}^{(u)}(A')$ is the probability of $u(\xi) \subset A'$.

We still have to prove the above-mentioned properties of $\mathfrak{F}^{(u)}$ and $\mathsf{P}^{(u)}$. They follow, however, from a single fact, namely:

LEMMA. *The sum, product, and difference of any pre-image sets $u^{-1}(A')$ are the pre-images of the corresponding sums, products, and differences of the original sets A'.*

The proof of this lemma is left for the reader.

Let A' and B' be two sets of $\mathfrak{F}^{(u)}$. Their pre-images A and B belong then to $\mathfrak{F}$. Since $\mathfrak{F}$ is a field, the sets AB, $A + B$, and $A - B$ also belong to $\mathfrak{F}$; but these sets are the pre-images of the sets $A'B'$, $A' + B'$, and $A' - B'$, which thus belong to $\mathfrak{F}^{(u)}$. This proves that $\mathfrak{F}^{(u)}$ is a field. In the same manner it can be shown that if $\mathfrak{F}$ is a Borel field, so is $\mathfrak{F}^{(u)}$.

Furthermore, it is clear that

$$\mathsf{P}^{(u)}(E') = \mathsf{P}\{u^{-1}(E')\} = \mathsf{P}(E) = 1.$$

That $\mathsf{P}^{(u)}$ is always non-negative, is self-evident. It remains only to be shown, therefore, that $\mathsf{P}^{(u)}$ is completely additive (cf. the end of § 1, Chap. II).

Let us assume that the sets A'_n, and therefore their pre-images $u^{-1}(A'_n)$, are disjoint. It follows that

$$\mathsf{P}^{(u)}(\sum_n A'_n) = \mathsf{P}\{u^{-1}(\sum_n A'_n)\} = \mathsf{P}\{\sum_n u^{-1}(A'_n)\}$$
$$= \sum_n \mathsf{P}\{u^{-1}(A'_n)\} = \sum_n \mathsf{P}^{(u)}(A'_n)$$

which proves the complete additivity of $\mathsf{P}^{(u)}$.

In conclusion let us also note the following. Let $u_1(\xi)$ be a function mapping E on E', and $u_2(\xi')$ be another function, mapping E' on E''. The product function $u_2u_1(\xi)$ maps E on E''. We shall now study the probability functions $\mathsf{P}^{(u_1)}(A')$ and $\mathsf{P}^{(u)}(A'')$ for the functions $u_1(\xi)$ and $u(\xi) = u_2u_1(\xi)$. It is easy to show that these two probability functions are connected by the following relation:

$$\mathsf{P}^{(u)}(A'') = \mathsf{P}^{(u_1)}\{u_2^{-1}(A'')\}. \tag{2}$$

§ 2. Definition of Random Variables and of Distribution Functions

DEFINITION. A real single-valued function $x(\xi)$, defined on the basic set E, is called a *random variable* if for each choice of a real number a the set $\{x < a\}$ of all ξ for which the inequality $x < a$ holds true, belongs to the system of sets $\mathfrak{F}$.

This function $x(\xi)$ maps the basic set E into the set R^1 of all real numbers. This function determines, as in § 1, a field $\mathfrak{F}^{(x)}$ of subsets of the set R^1. We may formulate our definition of random variable in this manner: A real function $x(\xi)$ is a random variable if and only if $\mathfrak{F}^{(x)}$ contains every interval of the form $(-\infty; a)$.

Since $\mathfrak{F}^{(x)}$ is a field, then along with the intervals $(-\infty; a)$ it contains all possible finite sums of half-open intervals $[a; b)$. If our field of probability is a Borel field, then $\mathfrak{F}$ and $\mathfrak{F}^{(x)}$ are Borel fields; *therefore, in this case $\mathfrak{F}^{(x)}$ contains all Borel sets of R^1.*

The probability function of a random variable we shall denote in the future by $\mathsf{P}^{(x)}(A')$. It is defined for all sets of the field $\mathfrak{F}^{(x)}$. In particular, for the most important case, the Borel field of probability, $\mathsf{P}^{(x)}$ is defined for all Borel sets of R^1.

DEFINITION. The function

$$F^{(x)}(a) = \mathsf{P}^{(x)}(-\infty, a) = \mathsf{P}\;\{x < a\},$$

where $-\infty$ and $+\infty$ are allowable values of a, is called the *distribution function of the random variable x.*

From the definition it follows at once that

$$F^{(x)}(-\infty) = 0,\; F^{(x)}(+\infty) = 1\;. \tag{1}$$

The probability of the realization of both inequalities $a \leqq x < b$, is obviously given by the formula

$$\mathsf{P}\{x \subset [a; b)\} = F^{(x)}(b) - F^{(x)}(a) \tag{2}$$

From this, we have, for $a < b$,

$$F^{(x)}(a) \leqq F^{(x)}(b)$$

which means that $F^{(x)}(a)$ is a *non-decreasing function.* Now let $a_1 < a_2 < \ldots < a_n < \ldots < b$; then

$$\mathfrak{D}_n\{x \subset [a_n; b)\} = 0$$

Therefore, in accordance with the continuity axiom,

$$F^{(x)}(b) - F^{(x)}(a_n) = \mathsf{P}\{x \subset [a_n, b)\}$$

approaches zero as $n \to +\infty$. From this it is clear that $F^{(x)}(a)$ *is continuous on the left.*

In an analogous way we can prove the formulae:

$$\lim F^{(x)}(a) = F^{(x)}(-\infty) = 0, \qquad a \to -\infty, \tag{3}$$

$$\lim F^{(x)}(a) = F^{(x)}(+\infty) = 1, \qquad a \to +\infty. \tag{4}$$

If the field of probability $(\mathfrak{F}, \mathsf{P})$ is a Borel field, the values of the probability function $\mathsf{P}^{(x)}(A)$ for all Borel sets A of R^1 are uniquely determined by knowledge of the distribution function

$F^{(x)}(a)$ (cf. § 3, III in Chap. II). Since our main interest lies in these values of $P^{(x)}(A)$, the distribution function plays a most significant role in all our future work.

If the distribution function $F^{(x)}(a)$ is differentiable, then we call its derivative with respect to a,

$$f^{(x)}(a) = \frac{d}{da} F^{(x)}(a) \,,$$

the *probability density* of x at the point a.

If also $F^{(x)}(a) = \int\limits_{-\infty}^{a} f^{(x)}(a) \, da$ for each a, then we may express the probability function $P^{(x)}(A)$ for each Borel set A in terms of $f^{(x)}(a)$ in the following manner:

$$P^{(x)}(A) = \int\limits_{A} f^{(x)}(a) \, da. \tag{5}$$

In this case we call the *distribution of x continuous*. And in the general case, we write, analogously

$$P^{(x)}(A) = \int\limits_{A} d F^{(x)}(a) . \tag{6}$$

All the concepts just introduced are capable of generalization for conditional probabilities. The set function

$$P_B^{(x)}(A) = P_B(x \subset A)$$

is the conditional probability function of x under hypothesis B. The non-decreasing function

$$F_B^{(x)}(a) = P_B(x < a)$$

is the corresponding distribution function, and, finally (in the case where $F_B^{(x)}(a)$ is differentiable)

$$f_B^{(x)}(a) = \frac{d}{da} F_B^{(x)}(a)$$

is the conditional probability density of x at the point a under hypothesis B.

§ 3. Multi-dimensional Distribution Functions

Let now n random variables $x_1, x_2, \ldots, x_n$ be given. The point $x = (x_1, x_2, \ldots, x_n)$ of the n-dimensional space R^n is a function of the elementary event ξ. Therefore, according to the general rules in § 1, we have a field $\mathfrak{F}^{(x_1, x_2, \ldots, x_n)}$ consisting of subsets of space R^n and a probability function $\mathsf{P}^{(x_1, x_2, \ldots, x_n)}(A')$ defined on $\mathfrak{F}'$. This probability function is called the n-dimensional *probability function of the random variables* $x_1, x_2, \ldots, x_n$.

As follows directly from the definition of a random variable, the field $\mathfrak{F}'$ contains, for each choice of i and a_i $(i = 1, 2, \ldots, n)$, the set of all points in R^n for which $x_i < a_i$. Therefore $\mathfrak{F}'$ also contains the intersection of the above sets, i.e. the set $L_{a_1 a_2 \ldots a_n}$ of all points of R^n for which all the inequalities $x_i < a_i$ hold $(i = 1, 2, \ldots, n)$[1].

If we now denote as the n-dimensional half-open interval

$$[a_1, a_2, \ldots, a_n; b_1, b_2, \ldots, b_n) ,$$

the set of all points in R^n, for which $a_i \leqq x_i < b_i$, then we see at once that each such interval belongs to the field $\mathfrak{F}'$ since

$$[a_1, a_2, \ldots, a_n; \quad b_1, b_2, \ldots, b_n)$$
$$= L_{b_1 b_2 \ldots b_n} - L_{a_1 b_2 \ldots b_n} - L_{b_1 a_2 b_3 \ldots b_n} - \cdots - L_{b_1 b_2 \ldots b_{n-1} a_n} .$$

The Borel extension of the system of all n-dimensional half-open intervals consists of all Borel sets in R^n. From this it follows that *in the case of a Borel field of probability, the field $\mathfrak{F}$ contains all the Borel sets in the space R^n*.

THEOREM : *In the case of a Borel field of probability each Borel function $x = f(x_1, x_2, \ldots, x_n)$ of a finite number of random variables $x_1, x_2, \ldots, x_n$ is also a random variable.*

All we need to prove this is to point out that the set of all points $(x_1, x_2, \ldots, x_n)$ in R^n for which $x = f(x_1, x_2, \ldots, x_n) < a$, is a Borel set. In particular, all finite sums and products of random variables are also random variables.

DEFINITION : The function

$$F^{(x_1, x_2, \ldots, x_n)}(a_1, a_2, \ldots, a_n) = \mathsf{P}^{(x_1, x_2, \ldots, x_n)}(L_{a_1 a_2 \ldots a_n})$$

is called the n-dimensional *distribution function* of the random variables $x_1, x_2, \ldots, x_n$.

As in the one-dimensional case, we prove that the n-dimensional distribution function $F^{(x_1, x_2, \ldots, x_n)}(a_1, a_2, \ldots, a_n)$ is non-decreasing and continuous on the left in each variable. In analogy to equations (3) and (4) in § 2, we here have

---

[1] The a_i may also assume the infinite values $\pm \infty$.

$$\lim_{a_i \to -\infty} F(a_1, a_2, \ldots, a_n) = F(a_1, \ldots, a_{i-1}, -\infty, a_{i+1}, \ldots, a_n) = 0, \quad (7)$$

$$\lim_{a_1 \to +\infty,\, a_2 \to +\infty, \ldots,\, a_n \to +\infty} F(a_1, a_2, \ldots, a_n) = F(+\infty, +\infty, \ldots, +\infty) = 1. \quad (8)$$

The distribution function $F^{(x_1 \, x_2 \, \ldots \, x_n)}$ gives directly the values of $\mathsf{P}^{(x_1, x_2, \ldots, x_n)}$ only for the special sets $L_{a_1 a_2 \ldots a_n}$. If our field, however, is a Borel field, then[2] $\mathsf{P}^{(x_1, x_2, \ldots, x_n)}$ *is uniquely determined for all Borel sets in R^n by knowledge of the distribution function* $F^{(x_1, x_2, \ldots, x_n)}$.

If there exists the derivative

$$f(a_1, a_2, \ldots, a_n) = \frac{\partial^n}{\partial a_1\, \partial a_2 \ldots \partial a_n} F^{(x_1, x_2, \ldots, x_n)}(a_1, a_2, \ldots, a_n)$$

we call this derivative the *n-dimensional probability density* of the random variables $x_1, x_2, \ldots, x_n$ at the point $a_1, a_2, \ldots, a_n$. If also for every point $(a_1, a_2, \ldots, a_n)$

$$F^{(x_1, x_2, \ldots, x_n)}(a_1 a_2 \ldots a_n) = \int_{-\infty}^{a_1} \int_{-\infty}^{a_2} \ldots \int_{-\infty}^{a_n} f(a_1, a_2, \ldots, a_n)\, da_1\, da_2 \ldots da_n,$$

then the distribution of $x_1, x_2, \ldots, x_n$ is called *continuous*. For every Borel set $A \subset R^n$, we have the equality

$$\mathsf{P}^{(x_1, x_2, \ldots, x_n)}(A) = \underset{A}{\int\int \ldots \int} f(a_1, a_2, \ldots, a_n)\, da_1\, da_2 \ldots da_n. \quad (9)$$

In closing this section we shall make one more remark about the relationships between the various probability functions and distribution functions.

Given the substitution

$$S = \begin{pmatrix} 1, & 2, & \ldots, & n \\ i_1, & i_2, & \ldots, & i_n \end{pmatrix},$$

and let r_S denote the transformation

$$x'_k = x_{i_k} \qquad (k = 1, 2, \ldots, n)$$

of space R^n into itself. It is then obvious that

$$\mathsf{P}^{(x_{i_1}, x_{i_2}, \ldots, x_{i_n})}(A) = \mathsf{P}^{(x_1, x_2, \ldots, x_n)}\{r_S^{-1}(A)\}. \quad (10)$$

Now let $x' = p_k(x)$ be the "projection" of the space R^n on the space R^k $(k < n)$, so that the point $(x_1, x_2, \ldots, x_n)$ is mapped onto the point $(x_1, x_2, \ldots, x_k)$. Then, as a result of Formula (2) in § 1,

---

[2] Cf. § 3, IV in the Second Chapter.

$$\mathsf{P}^{(x_1, x_2, \ldots, x_k)}(A) = \mathsf{P}^{(x_1, x_2, \ldots, x_n)}\{p_k^{-1}(A)\}. \tag{11}$$

For the corresponding distribution functions, we obtain from (10) and (11) the equations:

$$F^{(x_{i_1}, x_{i_2}, \ldots, x_{i_n})}(a_{i_1}, a_{i_2}, \ldots, a_{i_n}) = F^{(x_1, x_2, \ldots, x_n)}(a_1, a_2, \ldots, a_n), \tag{12}$$

$$F^{(x_1, x_2, \ldots, x_k)}(a_1, a_2, \ldots, a_k) = F^{(x_1, x_2, \ldots, x_n)}(a_1, \ldots, a_k, +\infty, \ldots, +\infty). \tag{13}$$

§ 4. Probabilities in Infinite-dimensional Spaces

In § 3 of the second chapter we have seen how to construct various fields of probability common in the theory of probability. We can imagine, however, interesting problems in which the elementary events are defined by means of an infinite number of coordinates. Let us take a set M of indices μ (indexing set) of arbitrary cardinality $\mathfrak{m}$. The totality of all systems

$$\xi = \{x_\mu\}$$

of real numbers x_μ, where μ runs through the entire set M, we shall call the space R^M (in order to define an element ξ in space R^M, we must put each element μ in set M in correspondence with a real number x_μ or, equivalently, assign a real single-valued function x_μ of the element μ, defined on M)[3]. If the set M consists of the first n natural numbers $1, 2, \ldots, n$, then R^M is the ordinary n-dimensional space R^n. If we choose for the set M all real numbers R^1, then the corresponding space $R^M = R^{R^1}$ will consist of all real functions

$$\xi(\mu) = x_\mu$$

of the real variable μ.

We now take the set R^M (with an arbitrary set M) as the basic set E. Let $\xi = \{x_\mu\}$ be an element in E; we shall denote by $p_{\mu_1 \mu_2 \ldots \mu_n}(\xi)$ the point $(x_{\mu_1}, x_{\mu_2}, \ldots, x_{\mu_n})$ of the n-dimensional space R^n. A subset A of E we shall call a *cylinder set* if it can be represented in the form

$$A = p_{\mu_1 \mu_2 \ldots \mu_n}^{-1}(A')$$

where A' is a subset of R^n. The class of all cylinder sets coincides, therefore, with the class of all sets which can be defined by relations of the form

---

[3] Cf. HAUSDORFF, *Mengenlehre*, 1927, p. 23.

$$f(x_{\mu_1}, x_{\mu_2}, \ldots, x_{\mu_n}) = 0 \quad . \tag{1}$$

In order to determine an arbitrary cylinder set $p_{\mu_1 \mu_2 \ldots \mu_n}(A')$ by such a relation, we need only take as f a function which equals 0 on A', but outside of A' equals unity.

A cylinder set is *a Borel cylinder set* if the corresponding set A' is a Borel set. *All Borel cylinder sets of the space R^M form a field, which we shall henceforth denote by $\mathfrak{F}^{M}$*[4].

The Borel extension of the field $\mathfrak{F}^M$ we shall denote, as always, by $B\mathfrak{F}^M$. Sets in $B\mathfrak{F}^M$ we shall call *Borel sets of the space R^M*.

Later on we shall give a method of constructing and operating with probability functions on $\mathfrak{F}^M$, and consequently, by means of the Extension Theorem, on $B\mathfrak{F}^M$ also. We obtain in this manner fields of probability sufficient for all purposes in the case that the set M is denumerable. We can therefore handle all questions touching upon a denumerable sequence of random variables. But if M is not denumerable, many simple and interesting subsets of R^M remain outside of $B\mathfrak{F}^M$. For example, the set of all elements ξ for which x_μ remains smaller than a fixed constant for all indices μ, does not belong to the system $B\mathfrak{F}^M$ if the set M is non-denumerable.

It is therefore desirable to try whenever possible to put each problem in such a form that the space of all elementary events ξ has only a denumerable set of coordinates.

Let a probability function $P(A)$ be defined on $\mathfrak{F}^M$. We may then regard every coordinate x_μ of the elementary event ξ as a random variable. In consequence, every finite group $(x_{\mu_1}, x_{\mu_2}, \ldots, x_{\mu_n})$ of these coordinates has an n-dimensional probability function $P_{\mu_1 \mu_2 \ldots \mu_n}(A)$ and a corresponding distribu-

---

[4] From the above it follows that Borel cylinder sets are Borel sets definable by relations of type (1). Now let A and B be two Borel cylinder sets defined by the relations

$$f(x_{\mu_1}, x_{\mu_2}, \ldots, x_{\mu_n}) = 0, \qquad g(x_{\lambda_1}, x_{\lambda_2}, \ldots, x_{\lambda_m}) = 0 \quad .$$

Then we can define the sets $A + B$, AB, and $A - B$ respectively by the relations

$$f \cdot g = 0,$$
$$f^2 + g^2 = 0,$$
$$f^2 + \omega(g) = 0,$$

where $\omega(x) = 0$ for $x \neq 0$ and $\omega(0) = 1$ If f and g are Borel functions, so also are $f \cdot g$, $f^2 + g^2$ and $f^2 + \omega(g)$; therefore, $A + B$, AB and $A - B$ are Borel cylinder sets. Thus we have shown that the system of sets $\mathfrak{F}^M$ is a field.

tion function $\quad F_{\mu_1\mu_2\ldots\mu_n}(a_1, a_2, \ldots, a_n)$. It is obvious that for every Borel cylinder set

$$A = p_{\mu_1\mu_2\ldots\mu_n}^{-1}(A'),$$

the following equation holds:

$$P(A) = P_{\mu_1\mu_2\ldots\mu_n}(A'),$$

where A' is a Borel set of R^n. In this manner, the probability function P is uniquely determined on the field $\mathfrak{F}^M$ of all cylinder sets by means of the values of all finite probability functions $P_{\mu_1\mu_2\ldots\mu_n}$ for all Borel sets of the corresponding spaces R^n. However, for Borel sets, the values of the probability functions $P_{\mu_1\mu_2\ldots\mu_n}$ are uniquely determined by means of the corresponding distribution functions. We have thus proved the following theorem:

The set of all finite-dimensional distribution functions $F_{\mu_1\mu_2\ldots\mu_n}$ uniquely determines the probability function $P(A)$ for all sets in $\mathfrak{F}^M$. If $P(A)$ is defined on $\mathfrak{F}^M$, then (according to the extension theorem) it is uniquely determined on $B\mathfrak{F}^M$ by the values of the distribution functions $F_{\mu_1\mu_2\ldots\mu_n}$.

We may now ask the following. Under what conditions does a system of distribution functions $F_{\mu_1\mu_2\ldots\mu_n}$ given *a priori* define a field of probability on $\mathfrak{F}^M$ (and, consequently, on $B\mathfrak{F}^M$)?

We must first note that every distribution function $F_{\mu_1\mu_2\ldots\mu_n}$ must satisfy the conditions given in § 3, III of the second chapter; indeed this is contained in the very concept of distribution function. Besides, as a result of formulas (13) and (14) in § 2, we have also the following relations:

$$F_{\mu_{i_1}\mu_{i_2}\ldots\mu_{i_n}}(a_{i_1}, a_{i_2}, \ldots, a_{i_n}) = F_{\mu_1\mu_2\ldots\mu_n}(a_1, a_2, \ldots, a_n),\qquad\textbf{(2)}$$

$$F_{\mu_1\mu_2\ldots\mu_k}(a_1, a_2, \ldots, a_k) = F_{\mu_1\mu_2\ldots\mu_n}(a_1, a_2, \ldots, a_k, +\infty, \ldots, +\infty),\textbf{(3)}$$

where $k < n$ and $\begin{pmatrix}1, & 2, & \ldots, & n \\ i_1, & i_2, & \ldots, & i_n\end{pmatrix}$ is an arbitrary permutation. These necessary conditions prove also to be sufficient, as will appear from the following theorem.

FUNDAMENTAL THEOREM: *Every system of distribution functions $F_{\mu_1\mu_2\ldots\mu_n}$, satisfying the conditions (2) and (3), defines a probability function $P(A)$ on $\mathfrak{F}^M$, which satisfies Axioms I - VI. This probability function $P(A)$ can be extended (by the extension theorem) to $B\mathfrak{F}^M$ also.*

Proof. Given the distribution functions $F_{\mu_1\mu_2\ldots\mu_n}$, satisfying the general conditions of Chap. II, § 3, III and also conditions (2) and (3). Every distribution function $F_{\mu_1\mu_2\ldots\mu_n}$ defines uniquely a corresponding probability function $\mathsf{P}_{\mu_1\mu_2\ldots\mu_n}$ for all Borel sets of R^n (cf. § 3). We shall deal in the future only with Borel sets of R^n and with Borel cylinder sets in E.

For every cylinder set

$$A = p^{-1}_{\mu_1\mu_2\ldots\mu_n}(A') \, ,$$

we set

$$\mathsf{P}(A) = \mathsf{P}_{\mu_1\mu_2\ldots\mu_n}(A') \, . \tag{4}$$

Since the same cylinder set A can be defined by various sets A', we must first show that formula (4) yields always the same value for $\mathsf{P}(A)$.

Let $(x_{\mu_1}, x_{\mu_2}, \ldots, x_{\mu_n})$ be a finite system of random variables x_μ. Proceeding from the probability function $\mathsf{P}_{\mu_1\mu_2\ldots\mu_n}$ of these random variables, we can, in accordance with the rules in § 3, define the probability function $\mathsf{P}_{\mu_{i_1}\mu_{i_2}\ldots\mu_{i_k}}$ of each subsystem $(x_{\mu_{i_1}}, x_{\mu_{i_2}}, \ldots, x_{\mu_{i_k}})$. From equations (2) and (3) it follows that this probability function defined according to § 3 is the same as the function $\mathsf{P}_{\mu_{i_1}\mu_{i_2}\ldots\mu_{i_k}}$ given *a priori*. We shall now suppose that the cylinder set A is defined by means of

$$A = p^{-1}_{\mu_{i_1}\mu_{i_2}\ldots\mu_{i_k}}(A')$$

and simultaneously by means of

$$A = p^{-1}_{\mu_{j_1}\mu_{j_2}\ldots\mu_{j_m}}(A'')$$

where all random variables x_{μ_i} and x_{μ_j} belong to the system $(x_{\mu_1}, x_{\mu_2}, \ldots, x_{\mu_n})$, which is obviously not an essential restriction. The conditions

$$\left(x_{\mu_{i_1}}, x_{\mu_{i_2}}, \ldots, x_{\mu_{i_k}}\right) \subset A'$$

and

$$\left(x_{\mu_{j_1}}, x_{\mu_{j_2}}, \ldots, x_{\mu_{j_m}}\right) \subset A''$$

are equivalent. Therefore

$$\mathsf{P}_{\mu_{i_1}\mu_{i_2}\ldots\mu_{i_k}}(A') = \mathsf{P}_{\mu_1\mu_2\ldots\mu_n}\left\{\left(x_{\mu_{i_1}}, x_{\mu_{i_2}}, \ldots, x_{\mu_{i_k}}\right) \subset A'\right\}$$

$$= \mathsf{P}_{\mu_1\mu_2\ldots\mu_n}\left\{\left(x_{\mu_{j_1}}, x_{\mu_{j_2}}, \ldots, x_{\mu_{j_m}}\right) \subset A''\right\} = \mathsf{P}_{\mu_{j_1}\mu_{j_2}\ldots\mu_{j_m}}(A'') \, ,$$

which proves our statement concerning the uniqueness of the definition of $\mathsf{P}(A)$.

Let us now prove that the field of probability $(\mathfrak{F}^M, \mathsf{P})$ satisfies all the Axioms I - VI. Axiom I requires merely that $\mathfrak{F}^M$ be a field. This fact has already been proven above. Moreover, for an arbitrary μ :

$$E = p_\mu^{-1}(R^1),$$
$$\mathsf{P}(E) = \mathsf{P}_\mu(R^1) = 1,$$

which proves that Axioms II and IV apply in this case. Finally, from the definition of $\mathsf{P}(A)$ it follows at once that $\mathsf{P}(A)$ is non-negative (Axiom III).

It is only slightly more complicated to prove that Axiom V is also satisfied. In order to do so, we investigate two cylinder sets

$$A = p_{\mu_{i_1}\mu_{i_2}\ldots\mu_{i_k}}^{-1}(A')$$

and

$$B = p_{\mu_{j_1}\mu_{j_2}\ldots\mu_{j_m}}^{-1}(B').$$

We shall assume that all variables x_{μ_i} and x_{μ_j} belong to one inclusive finite system $(x_{\mu_1}, x_{\mu_2}, \ldots, x_{\mu_n})$. If the sets A and B do not intersect, the relations

$$\left(x_{\mu_{i_1}}, x_{\mu_{i_2}}, \ldots, x_{\mu_{i_k}}\right) \subset A'$$

and

$$\left(x_{\mu_{j_1}}, x_{\mu_{j_2}}, \ldots, x_{\mu_{j_k}}\right) \subset B'$$

are incompatible. Therefore

$$\mathsf{P}(A + B) = \mathsf{P}_{\mu_1\mu_2\ldots\mu_n}\left\{\left(x_{\mu_{i_1}}, x_{\mu_{i_2}}, \ldots, x_{\mu_{i_k}}\right) \subset A'\right.$$
$$\text{or} \quad \left(x_{\mu_{j_1}}, x_{\mu_{j_2}}, \ldots, x_{\mu_{j_m}}\right) \subset B'\right\}$$
$$= \mathsf{P}_{\mu_1\mu_2\ldots\mu_n}\left\{\left(x_{\mu_{i_1}}, x_{\mu_{i_2}}, \ldots, x_{\mu_{i_k}}\right) \subset A'\right\}$$
$$+ \mathsf{P}_{\mu_1\mu_2\ldots\mu_n}\left\{\left(x_{\mu_{j_1}}, x_{\mu_{j_2}}, \ldots, x_{\mu_{j_m}}\right) \subset B'\right\} = \mathsf{P}(A) + \mathsf{P}(B),$$

which concludes our proof.

Only Axiom VI remains. Let

$$A_1 \supset A_2 \supset \cdots \supset A_n \supset \cdots$$

be a decreasing sequence of cylinder sets satisfying the condition

$$\lim \mathsf{P}(A_n) = L > 0.$$

We shall prove that the product of all sets A_n is not empty. We may assume, without essentially restricting the problem, that in the definition of the first n cylinder sets A_k, only the first n co-ordinates x_{μ_k} in the sequence

$$x_{\mu_1}, x_{\mu_2}, \ldots, x_{\mu_n}, \ldots$$

occur, i.e.

$$A_n = p^{-1}_{\mu_1 \mu_2 \ldots \mu_n}(B_n).$$

For brevity we set

$$P_{\mu_1 \mu_2 \ldots \mu_n}(B) = P_n(B);$$

then, obviously

$$P_n(B_n) = P(A_n) \geqq L > 0.$$

In each set B_n it is possible to find a closed bounded set U_n such that

$$P_n(B_n - U_n) \leqq \frac{\varepsilon}{2^n}.$$

From this inequality we have for the set

$$V_n = p^{-1}_{\mu_1 \mu_2 \ldots \mu_n}(U_n)$$

the inequality

$$P(A_n - V_n) \leqq \frac{\varepsilon}{2^n}. \tag{5}$$

Let, morever,

$$W_n = V_1 V_2 \ldots V_n.$$

From (5) it follows that

$$P(A_n - W_n) \leqq \varepsilon.$$

Since $W_n \subset V_n \subset A_n$, it follows that

$$P(W_n) \geqq P(A_n) - \varepsilon \geqq L - \varepsilon.$$

If ε is sufficiently small, $P(W_n) > 0$ and W_n is not empty. We shall now choose in each set W_n a point $\xi^{(n)}$ with the coordinates $x^{(n)}_\mu$. Every point $\xi^{(n+p)}$, $p = 0, 1, 2, \ldots$, belongs to the set V_n; therefore

$$\left(x^{(n+p)}_{\mu_1}, x^{(n+p)}_{\mu_2}, \ldots, x^{(n+p)}_{\mu_n}\right) = p^{-1}_{\mu_1 \mu_2 \ldots \mu_n}(\xi^{(n+p)}) \subset U_n.$$

Since the sets U_n are bounded we may (by the diagonal method) choose from the sequence $\{\xi^{(n)}\}$ a subsequence

$$\xi^{(n_1)}, \; \xi^{(n_2)}, \; \ldots, \; \xi^{(n_i)}, \; \ldots$$

for which the corresponding coordinates $x^{(n_i)}_{\mu_k}$ tend for any k to a definite limit x_k. Let, finally, ξ be a point in set E with the coordinates

$$x_{\mu_k} = x_k ,$$
$$x_\mu = 0, \quad \mu \neq \mu_k . \qquad\qquad k = 1, 2, 3, \ldots$$

As the limit of the sequence $(x_1^{(n_i)}, x_2^{(n_i)}, \ldots, x_k^{(n_i)})$, $i = 1, 2, 3, \ldots$, the point $(x_1, x_2, \ldots, x_k)$ belongs to the set U_k. Therefore, ξ belongs to

$$A_k \subset V_k = p_{\mu_1 \mu_2 \ldots \mu_k}^{-1}(U_k)$$

for any k and therefore to the product

$$A = \mathop{\mathfrak{D}}_{k} A_k .$$

§ 5. Equivalent Random Variables; Various Kinds of Convergence

Starting with this paragraph, we deal exclusively with Borel fields of probability. As we have already explained in § 2 of the second chapter, this does not constitute any essential restriction on our investigations.

Two random variables x and y are called *equivalent*, if the probability of the relation $x \neq y$ is equal to zero. It is obvious that two equivalent random variables have the same probability function:

$$\mathsf{P}^{(x)}(A) = \mathsf{P}^{(y)}(A).$$

Therefore, the distribution functions $F^{(x)}$ and $F^{(y)}$ are also identical. In many problems in the theory of probability we may substitute for any random variable any equivalent variable.

Now let

$$x_1, x_2, \ldots, x_n, \ldots \qquad\qquad (1)$$

be a sequence of random variables. Let us study the set A of all elementary events ξ for which the sequence (1) converges. If we denote by $A_{np}^{(m)}$ the sets of ξ for which all the following inequalities hold

$$|x_{n+k} - x_n| < \frac{1}{m} \qquad\qquad k = 1, 2, \ldots, p$$

then we obtain at once

$$A = \mathop{\mathfrak{D}}_{m} \mathop{\mathfrak{S}}_{n} \mathop{\mathfrak{D}}_{p} A_{np}^{(m)} . \qquad\qquad (2)$$

According to § 3, the set $A_{np}^{(m)}$ always belongs to the field $\mathfrak{F}$. The relation (2) shows that A, too, belongs to $\mathfrak{F}$. *We may, therefore, speak of the probability of convergence of a sequence of random variables, for it always has a perfectly definite meaning.*

Now let the probability $\mathsf{P}(A)$ of the convergence set A be equal to *unity*. We may then state that the sequence (1) converges with the probability one to a random variable x, where the random variable x is uniquely defined except for equivalence. To determine such a random variable we set

$$x = \lim x_n \qquad\qquad n \to \infty$$

on A, and $x = 0$ outside of A. We have to show that x is a random variable, in other words, that the set $A(a)$ of the elements ξ for which $x < a$, belongs to $\mathfrak{F}$. But

$$A(a) = A \mathop{\mathfrak{S}}_{n} \mathop{\mathfrak{D}}_{p} \{x_{n+p} < a\}$$

in case $a \leq 0$, and

$$A(a) = A \mathop{\mathfrak{S}}_{n} \mathop{\mathfrak{D}}_{p} \{x_{n+p} < a\} + \overline{A}$$

in the opposite case, from which our statement follows at once.

If the probability of convergence of the sequence (1) to x equals one, then we say that the sequence (1) converges *almost surely* to x. However, for the theory of probability, another conception of convergence is possibly more important.

DEFINITION. The sequence $x_1, x_2, \ldots, x_n, \ldots$ of random variables *converges in probability* (converge en probabilité) to the random variable x, if for any $\varepsilon > 0$, the probability

$$P\{|x_n - x| > \varepsilon\}$$

tends toward zero as $n \to \infty$ [5].

I. *If the sequence* (1) *converges in probability to x and also to x', then x and x' are equivalent.* In fact

$$P\left\{|x - x'| > \tfrac{1}{m}\right\} \leq P\left\{|x_n - x| > \tfrac{1}{2m}\right\} + P\left\{|x_n - x'| > \tfrac{1}{2m}\right\};$$

since the last probabilities are as small as we please for a sufficiently large n it follows that

$$P\left\{|x - x'| > \tfrac{1}{m}\right\} = 0$$

and we obtain at once that

$$P\{x \neq x'\} \leq \sum_m P\left\{|x - x'| > \tfrac{1}{m}\right\} = 0.$$

II. *If the sequence* (1) *almost surely converges to x, then it*

---

[5] This concept is due to Bernoulli; its completely general treatment was introduced by E. E. Slutsky (see [1]).

also converges to x in probability. Let A be the convergence set of the sequence (1); then

$$1 = \mathsf{P}(A) \leqq \lim_{n \to \infty} \mathsf{P}\{|x_{n+p} - x| < \varepsilon, p = 0, 1, 2, \ldots\} \leqq \lim_{n \to \infty} \mathsf{P}\{|x_n - x| < \varepsilon\},$$

from which the convergence in probability follows.

III. *For the convergence in probability of the sequence* (1) *the following condition is both necessary and sufficient: For any* $\xi > 0$ *there exists an n such that, for every p > 0, the following inequality holds*:

$$\mathsf{P}\{|x_{n+p} - x_n| > \varepsilon\} < \varepsilon \; .$$

Let $F_1(a)$, $F_2(a)$, $\ldots$, $F_n(a)$, $\ldots$, $F(a)$ be the distribution functions of the random variables x_1, x_2, $\ldots$, x_n, $\ldots$, x. If the sequence x_n converges in probability to x, the distribution function $F(a)$ is uniquely determined by knowledge of the functions $F_n(a)$. We have, in fact,

THEOREM: *If the sequence* x_1, x_2, $\ldots$, x_n, $\ldots$ *converges in probability to x, the corresponding sequence of distribution functions* $F_n(a)$ *converges at each point of continuity of* $F(a)$ *to the distribution function* $F(a)$ *of x.*

That $F(a)$ is really determined by the $F_n(a)$ follows from the fact that $F(a)$, being a monotone function, continuous on the left, is uniquely determined by its values at the points of continuity[6]. To prove the theorem we assume that F is continuous at the point a. Let $a' < a$; then in case $x < a'$, $x_n \geqq a$ it is necessary that $|x_n - x| > a - a'$. Therefore

$$\lim \mathsf{P}(x < a', x_n \geqq a) = 0 \, ,$$

$$F(a') = \mathsf{P}(x < a') \leqq \mathsf{P}(x_n < a) + \mathsf{P}(x < a', x_n \geqq a) = F_n(a) + \mathsf{P}(x < a', x_n \geqq a) \, ,$$

$$F(a') \leqq \lim \inf F_n(a) + \lim \mathsf{P}(x < a', x_n \geqq a) \, ,$$

$$F(a') \leqq \lim \inf F_n(a) \, . \tag{3}$$

In an analogous manner, we can prove that from $a'' > a$ there follows the relation

$$F(a'') \geqq \lim \sup F_n(a) \, . \tag{4}$$

---

[6] In fact, it has at most only a countable set of discontinuities (see LEBESGUE, *Leçons sur l'intégration*, 1928, p. 50. Therefore, the points of continuity are everywhere dense, and the value of the function $F(a)$ at a point of discontinuity is determined as the limit of its values at the points of continuity on its left.

Since $F(a')$ and $F(a'')$ converge to $F(a)$ for $a' \rightarrow a$ and $a'' \rightarrow a$, it follows from (3) and (4) that

$$\lim F_n(a) = F(a),$$

which proves our theorem.

Chapter IV

MATHEMATICAL EXPECTATIONS [1]

§ 1. Abstract Lebesgue Integrals

Let x be a random variable and A a set of $\mathfrak{F}$. Let us form, for a positive λ, the sum

$$S_\lambda = \sum_{k=-\infty}^{k=+\infty} k\lambda\, \mathsf{P}\{k\lambda \leqq x < (k+1)\lambda, \, \xi \subset A\}. \qquad (1)$$

If this series converges absolutely for every λ, then as $\lambda \to 0$, S_λ tends toward a definite limit, which is by definition the integral

$$\int_A x\mathsf{P}(dE) . \qquad (2)$$

In this abstract form the concept of an integral was introduced by Fréchet[2]; it is indispensable for the theory of probability. (The reader will see in the following paragraphs that the usual definition for the conditional mathematical expectation of the variable x under hypothesis A coincides with the definition of the integral (2) except for a constant factor.)

We shall give here a brief survey of the most important properties of the integrals of form (2). The reader will find their proofs in every textbook on real variables, although the proofs are usually carried out only in the case where $\mathsf{P}(A)$ is the Lebesgue measure of sets in R^n. The extension of these proofs to the general case does not entail any new mathematical problem; for the most part they remain word for word the same.

I. If a random variable x is integrable on A, then it is integrable on each subset A' of A belonging to $\mathfrak{F}$.

II. If x is integrable on A and A is decomposed into no

---

[1] As was stated in § 5 of the third chapter, we are considering in this, as well as in the following chapters, *Borel fields of probability* only.

[2] FRÉCHET, *Sur l'intégrale d'une functionnelle étendue à un ensemble abstrait*, Bull. Soc. Math. France v. 43, 1915, p. 248.

more than a countable number of non-intersecting sets A_n of $\mathfrak{F}$, then

$$\int_A x\,\mathsf{P}\,(dE) = \sum_n \int_{A_n} x\,\mathsf{P}\,(dE)\,.$$

III. If x is integrable, $|\,x\,|$ is also integrable, and in that case

$$\left|\int_A x\,\mathsf{P}\,(dE)\right| \leqq \int_A |x|\,\mathsf{P}\,(dE)\,.$$

IV. If in each event ξ, the inequalities $0 \leqq y \leqq x$ hold, then along with x, y is also integrable[3], and in that case

$$\int_A y\,\mathsf{P}\,(dE) \leqq \int_A x\,\mathsf{P}\,(dE)\,.$$

V. If $m \leqq x \leqq M$ where m and M are two constants, then

$$m\,\mathsf{P}\,(A) \leqq \int_A x\,\mathsf{P}\,(dE) \leqq M\,\mathsf{P}\,(A)\,.$$

VI. If x and y are integrable, and K and L are two real constants, then $Kx + Ly$ is also integrable, and in this case

$$\int_A (Kx + Ly)\,\mathsf{P}\,(dE) = K\int_A x\,\mathsf{P}\,(dE) + L\int_A y\,\mathsf{P}\,(dE)\,.$$

VII. If the series

$$\sum_n \int_A |x_n|\,\mathsf{P}\,(dE)$$

converges, then the series

$$\sum_n x_n = x$$

converges at each point of set A with the exception of a certain set B for which $\mathsf{P}\,(B) = 0$. If we set $x = 0$ everywhere except on $A - B$, then

$$\int_A x\,\mathsf{P}\,(dE) = \sum_n \int_A x_n\,\mathsf{P}\,(dE)\,.$$

VIII. If x and y are equivalent $(\mathsf{P}\,\{x \neq y\} = 0)$, then for every set A of $\mathfrak{F}$

$$\int_A x\,\mathsf{P}\,(dE) = \int_A y\,\mathsf{P}\,(dE)\,. \tag{3}$$

---

[3] It is assumed that y is a random variable, i.e., in the terminology of the general theory of integration, measurable with respect to $\mathfrak{F}$.

IX. If (3) holds for every set A of $\mathfrak{F}$, then x and y are equivalent.

From the foregoing definition of an integral we also obtain the following property, which is not found in the usual Lebesgue theory.

X. Let $P_1(A)$ and $P_2(A)$ be two probability functions defined on the same field $\mathfrak{F}$, $P(A) = P_1(A) + P_2(A)$, and let x be integrable on A relative to $P_1(A)$ and $P_2(A)$. Then

$$\int_A x P(dE) = \int_A x P_1(dE) + \int_A x P_2(dE).$$

XI. Every bounded random variable is integrable.

§ 2. Absolute and Conditional Mathematical Expectations

Let x be a random variable. The integral

$$E(x) = \int_E x P(dE)$$

is called in the theory of probability the *mathematical expectation* of the variable x. From the properties III, IV, V, VI, VII, VIII, XI, it follows that

    I. $|E(x)| \leqq E(|x|)$;

    II. $E(y) \leqq E(x)$ if $0 \leqq y \leqq x$ everywhere ;

    III. inf $(x) \leqq E(x) \leqq$ sup (x) ;

    IV. $E(Kx + Ly) = KE(x) + LE(y)$;

    V. $E\left(\sum_n x_n\right) = \sum_n E(x_n)$,   if the series $\sum_n E(|x_n|)$ converges ;

    VI. If x and y are equivalent then

$$E(x) = E(y).$$

VII. Every bounded random variable has a mathematical expectation.

From the definition of the integral, we have

$$E(x) = \lim \sum_{k=-\infty}^{k=+\infty} km\, P\{km \leqq x < (k+1)\,m\}$$

$$= \lim \sum_{k=-\infty}^{k=+\infty} km\{F((k+1)\,m) - F(km)\}.$$

The second line is nothing more than the usual definition of the Stieltjes integral

$$\int_{-\infty}^{+\infty} a \, dF^{(x)}(a) = \mathsf{E}(x).\tag{1}$$

Formula (1) may therefore serve as a definition of the mathematical expectation $\mathsf{E}(x)$.

Now let u be a function of the elementary event ξ, and x be a random variable defined as a single-valued function $x = x(u)$ of u. Then

$$\mathsf{P}\{km \leqq x < (k+1)\,m\} = \mathsf{P}^{(u)}\{km \leqq x(u) < (k+1)\,m\},$$

where $\mathsf{P}^{(u)}(A)$ is the probability function of u. It then follows from the definition of the integral that

$$\int_E x \, \mathsf{P}\,(dE) = \int_{E^{(u)}} x \, \mathsf{P}^{(u)}(dE^{(u)})$$

and, therefore,

$$\mathsf{E}(x) = \int_{E^{(u)}} x(u) \, \mathsf{P}^{(u)}(dE^{(u)})\tag{2}$$

where $E^{(u)}$ denotes the set of all possible values of u.

In particular, when u itself is a random variable we have

$$\mathsf{E}(x) = \int_E x \, \mathsf{P}\,(dE) = \int_{R^1} x(u) \, \mathsf{P}^{(u)}(dR^1) = \int_{-\infty}^{+\infty} x(a) \, dF^{(u)}(a).\tag{3}$$

When $x(u)$ is continuous, the last integral in (3) is the ordinary Stieltjes integral. We must note, however, that the integral

$$\int_{-\infty}^{+\infty} x(a) \, dF^{(u)}(a)$$

can exist even when the mathematical expectation $\mathsf{E}(x)$ does not. For the existence of $\mathsf{E}(x)$, it is necessary and sufficient that the integral

$$\int_{-\infty}^{+\infty} |x(a)| \, dF^{(u)}(a)$$

be finite[4].

If u is a point $(u_1, u_2, \ldots, u_n)$ of the space R^n, then as a result of (2):

---

[4] Cf. V. Glivenko, *Sur les valeurs probables de fonctions*, Rend. Accad. Lincei v. 8, 1928, pp. 480-483.

$$\mathsf{E}(x) = \int\!\!\int \cdots \int_{R^n} x(u_1, u_2, \ldots, u_n)\, \mathsf{P}^{(u_1, u_2, \ldots, u_n)}(d\,R^n)\,. \qquad (4)$$

We have already seen that the conditional probability $\mathsf{P}_B(A)$ possesses all the properties of a probability function. The corresponding integral

$$\mathsf{E}_B(x) = \int_E x\, \mathsf{P}_B(dE) \qquad (5)$$

we call *the conditional mathematical expectation of the random variable x with respect to the event B.* Since

$$\mathsf{P}_B(\bar{B}) = 0\,, \qquad \int_{\bar{B}} x\, \mathsf{P}_B(dE) = 0\,,$$

we obtain from (5) the equation

$$\mathsf{E}_B(x) = \int_E x\, \mathsf{P}_B(dE) = \int_B x\, \mathsf{P}_B(dE) + \int_{\bar{B}} x\, \mathsf{P}_B(dE) = \int_B x\, \mathsf{P}_B(dE)\,.$$

We recall that in case $A \subset B$,

$$\mathsf{P}_B(A) = \frac{\mathsf{P}(AB)}{\mathsf{P}(B)} = \frac{\mathsf{P}(A)}{\mathsf{P}(B)}\,;$$

we thus obtain

$$\mathsf{E}_B(x) = \frac{1}{\mathsf{P}(B)} \int_B x\, \mathsf{P}(dE)\,, \qquad (6)$$

$$\int_B x\, \mathsf{P}(dE) = \mathsf{P}(B) \cdot \mathsf{E}_B(x)\,. \qquad (7)$$

From (6) and the equality

$$\int_{A+B} x\, \mathsf{P}(dE) = \int_A x\, \mathsf{P}(dE) + \int_B x\, \mathsf{P}(dE)$$

we obtain at last

$$\mathsf{E}_{A+B}(x) = \frac{\mathsf{P}(A)\, \mathsf{E}_A(x) + \mathsf{P}(B)\, \mathsf{E}_B(x)}{\mathsf{P}(A+B)} \qquad (8)$$

and, in particular, we have the formula

$$\mathsf{E}(x) = \mathsf{P}(A)\, \mathsf{E}_A(x) + \mathsf{P}(\bar{A})\, \mathsf{E}_{\bar{A}}(x)\,. \qquad (9)$$

§ 3. The Tchebycheff Inequality

Let $f(x)$ be a non-negative function of a real argument x, which for $x \geqq a$ never becomes smaller than $b > 0$. Then for any random variable x

$$P(x \geqq a) \leqq \frac{E\{f(x)\}}{b}, \qquad (1)$$

provided the mathematical expectation $E\{f(x)\}$ exists. For,

$$E\{f(x)\} = \int_E f(x)\, P(dE) \geqq \int_{\{x \geqq a\}} f(x)\, P(dE) \geqq b\, P(x \geqq a),$$

from which (1) follows at once.

For example, for every positive c,

$$P(x \geqq a) \leqq \frac{E(e^{cx})}{e^{ca}}. \qquad (2)$$

Now let $f(x)$ be non-negative, even, and, for positive x, non-decreasing. Then for every random variable x and for any choice of the constant $a > 0$ the following inequality holds

$$P(|x| \geqq a) \leqq \frac{E\{f(x)\}}{f(a)}. \qquad (3)$$

In particular,

$$P(|x - E(x)| \geqq a) \leqq \frac{E f\{x - E(x)\}}{f(a)}. \qquad (4)$$

Especially important is the case $f(x) = x^2$. We then obtain from (3) and (4)

$$P(|x| \geqq a) \leqq \frac{E(x^2)}{a^2}, \qquad (5)$$

$$P(|x - E(x)| \geqq a) \leqq \frac{E\{x - E(x)\}^2}{a^2} = \frac{\sigma^2(x)}{a^2}, \qquad (6)$$

where

$$\sigma^2(x) = E\{x - E(x)\}^2$$

is called the *variance* of the variable x. It is easy to calculate that

$$\sigma^2(x) = E(x^2) - \{E(x)\}^2.$$

If $f(x)$ is bounded:

$$|f(x)| \leqq K,$$

then a lower bound for $P(|x| \geqq a)$ can be found. For

$$E(f(x)) = \int_E f(x) \, P(dE) = \int_{\{|x|<a\}} f(x) \, P(dE) + \int_{\{|x|\geq a\}} f(x) \, P(dE)$$

$$\leq f(a) \, P(|x| < a) + K P(|x| \geq a) \leq f(a) + K P(|x| \geq a)$$

and therefore

$$P(|x| \geq a) \geq \frac{E\{f(x)\} - f(a)}{K}. \tag{7}$$

If instead of $f(x)$ the random variable x itself is bounded,

$$|x| \leq M,$$

then $f(x) \leq f(M)$, and instead of (7), we have the formula

$$P(|x| \geq a) \geq \frac{E(f(x)) - f(a)}{f(M)}. \tag{8}$$

In the case $f(x) = x^2$, we have from (8)

$$P(|x| \geq a) \geq \frac{E(x^2) - a^2}{M^2}. \tag{9}$$

§ 4. Some Criteria for Convergence

Let

$$x_1, x_2, \ldots, x_n, \ldots \tag{1}$$

be a sequence of random variables and $f(x)$ be a non-negative, even, and for positive x a monotonically increasing function[5]. Then the following theorems are true:

I. In order that the sequence (1) converge in probability the following condition is sufficient: For each $\varepsilon > 0$ there exists an n such that for every $p > 0$, the following inequality holds:

$$E\{f(x_{n+p} - x_n)\} < \varepsilon. \tag{2}$$

II. In order that the sequence (1) converge in probability to the random variable x, the following condition is sufficient:

$$\lim_{n \to +\infty} E\{f(x_n - x)\} = 0. \tag{3}$$

III. If $f(x)$ is bounded and continuous and $f(0) = 0$, then conditions I and II are also necessary.

IV. If $f(x)$ is continuous, $f(0) = 0$, and the totality of all $x_1, x_2, \ldots, x_n, \ldots, x$ is bounded, then conditions I and II are also necessary.

---

[5] Therefore $f(x) > 0$ if $x \neq 0$.

From II and IV, we obtain in particular

V. In order that sequence (1) converge in probability to x, it is sufficient that

$$\lim \mathsf{E}(x_n - x)^2 = 0 \ . \tag{4}$$

If also the totality of all $x_1, x_2, \ldots, x_n, \ldots, x$ is bounded, then the condition is also necessary.

For proofs of I - IV see Slutsky [1] and Fréchet [1]. However, these theorems follow almost immediately from formulas (3) and (8) of the preceding section.

§ 5. Differentiation and Integration of Mathematical Expectations with Respect to a Parameter

Let us put each elementary event ξ into correspondence with a definite real function $x(t)$ of a real variable t. We say that $x(t)$ is a *random function* if for every fixed t, the variable $x(t)$ is a random variable. The question now arises, under what conditions can the mathematical expectation sign be interchanged with the integration and differentiation signs. The two following theorems, though they do not exhaust the problem, can nevertheless give a satisfactory answer to this question in many simple cases.

THEOREM I: *If the mathematical expectation* $\mathsf{E}[x(t)]$ *is finite for any t, and $x(t)$ is always differentiable for any t, while the derivative $x'(t)$ of $x(t)$ with respect to t is always less in absolute value than some constant M, then*

$$\frac{d}{dt} \mathsf{E}(x(t)) = \mathsf{E}(x'(t)) \ .$$

THEOREM II: *If $x(t)$ always remains less, in absolute value, than some constant K and is integrable in the Riemann sense, then*

$$\int_a^b \mathsf{E}(x(t))\, dt = \mathsf{E}\left[\int_a^b x(t)\, dt\right],$$

provided $\mathsf{E}[x(t)]$ *is integrable in the Riemann sense.*

Proof of Theorem I. Let us first note that $x'(t)$ as the limit of the random variables

$$\frac{x(t+h) - x(t)}{h} \qquad h = 1, \frac{1}{2}, \ldots, \frac{1}{n}, \ldots$$

is also a random variable. Since $x'(t)$ is bounded, the mathematical expectation $\mathsf{E}[x'(t)]$ exists (Property VII of mathematical expectation, in § 2).   Let us choose a **fixed** t and denote by A the event

$$\left| \frac{x(t+h)-x(t)}{h} - x'(t) \right| > \varepsilon .$$

The probability $\mathsf{P}(A)$ tends to zero as $h \to 0$ for every $\varepsilon > 0$. Since

$$\left| \frac{x(t+h)-x(t)}{h} \right| \leq M , \qquad |x(t)| \leq M$$

holds everywhere, and moreover in the case $\bar{A}$

$$\left| \frac{x(t+h)-x(t)}{h} - x'(t) \right| \leq \varepsilon ,$$

then

$$\left| \frac{\mathsf{E}x(t+h)-\mathsf{E}x(t)}{h} - \mathsf{E}x'(t) \right| \leq \mathsf{E}\left| \frac{x(t+h)-x(t)}{h} - x'(t) \right|$$

$$= \mathsf{P}(A)\,\mathsf{E}_A \left| \frac{x(t+h)-x(t)}{h} - x'(t) \right| + \mathsf{P}(\bar{A})\,\mathsf{E}_{\bar{A}} \left| \frac{x(t+h)-x(t)}{h} - x'(t) \right|$$

$$\leq 2M\,\mathsf{P}(A) + \varepsilon .$$

We may choose the $\varepsilon > 0$ arbitrarily, and $\mathsf{P}(A)$ is arbitrarily small for any sufficiently small h. Therefore

$$\frac{d}{dt}\mathsf{E}x(t) = \lim_{h \to 0} \frac{\mathsf{E}x(t+h)-\mathsf{E}x(t)}{h} = \mathsf{E}x'(t) ,$$

which was to be proved.

Proof of Theorem II. Let

$$S_n = \frac{1}{h} \sum_{k=1}^{k=n} x(t+kh) , \qquad h = \frac{b-a}{n} .$$

Since S_n converges to $J = \int_a^b x(t)\, dt$, we can choose for any $\varepsilon > 0$ an N such that from $n \geq N$ there follows the inequality

$$\mathsf{P}(A) = \mathsf{P}\{|S_k - J| > \varepsilon\} < \varepsilon .$$

If we set

$$S_n^* = \frac{1}{h} \sum_{k=1}^{k=n} \mathsf{E}x(t+kh) = \mathsf{E}(S_n) ,$$

then

$$|S_n^* - \mathsf{E}(J)| = |\mathsf{E}(S_n - J)| \leq \mathsf{E}|S_n - J|$$

$$= \mathsf{P}(A)\,\mathsf{E}_A|S_n - J| + \mathsf{P}(\bar{A})\,\mathsf{E}_{\bar{A}}|S_n - J|_i \leq 2K\,\mathsf{P}(A) + \varepsilon \leq (2K+1)\,\varepsilon .$$

Therefore, S_n^* converges to $\mathsf{E}(J)$, from which results the equation

$$\int_a^b \mathsf{E}\,x(t)\,dt = \lim S_n^* = \mathsf{E}(J).$$

Theorem II can easily be generalized for double and triple and higher order multiple integrals. We shall give an application of this theorem to one example in geometric probability. Let G be a measurable region of the plane whose shape depends on chance; in other words, let us assign to every elementary event ξ of a field of probability a definite measurable plane region G. We shall denote by J the area of the region G, and by $\mathsf{P}(x, y)$ the probability that the point (x, y) belongs to the region G. Then

$$\mathsf{E}(J) = \int\int \mathsf{P}(x,y)\,dx\,dy.$$

To prove this it is sufficient to note that

$$J = \int\int f(x,y)\,dx\,dy,$$
$$\mathsf{P}(x,y) = \mathsf{E}f(x,y),$$

where $f(x, y)$ is the characteristic function of the region G ($f(x, y) = 1$ on G and $f(x, y) = 0$ outside of G)[6].

---

[6] Cf. A. KOLMOGOROV and M. LEONTOVICH, *Zur Berechnung der mittleren Brownschen Fläche*, Physik. Zeitschr. d. Sovietunion, v. 4, 1933.

Chapter V

CONDITIONAL PROBABILITIES AND MATHEMATICAL EXPECTATIONS

§ 1. Conditional Probabilities

In § 6, Chapter I, we defined the conditional probability, $P_{\mathfrak{A}}(B)$, of the event B with respect to trial $\mathfrak{A}$. It was there assumed that $\mathfrak{A}$ allows of only a finite number of different possible results. We can, however, define $P_{\mathfrak{A}}(B)$ also for the case of an $\mathfrak{A}$ with an infinite set of possible results, i.e. the case in which the set E is partitioned into an infinite number of non-intersecting subsets. In particular, we obtain such a partitioning if we consider an arbitrary function u of ξ and define as elements of the partition $\mathfrak{A}_u$ the sets $u = $ constant. The conditional probability $P_{\mathfrak{A}_u}(B)$ we also denote by $P_u(B)$. Any partitioning $\mathfrak{A}$ of the set E can be defined as the partitioning $\mathfrak{A}_u$ which is "induced" by a function u of ξ, if one assigns to every ξ, as $u(\xi)$, that set of the partitioning $\mathfrak{A}$ of E which contains ξ.

Two functions u and u' of ξ determine the same partitioning $\mathfrak{A}_u = \mathfrak{A}_{u'}$ of the set E if and only if there exists a one-to-one correspondence $u' = f(u)$ between their domains $\mathfrak{F}^{(u)}$ and $\mathfrak{F}^{(u')}$ such that $u'(\xi)$ is identical with $fu(\xi)$. The reader can easily show that the random variables $P_u(B)$ and $P_{u'}(B)$, defined below, are in this case the same. They are thus determined, in fact, by the partition $\mathfrak{A}_u = \mathfrak{A}_{u'}$ itself.

To define $P_u(B)$ we may use the following equation:

$$P_{\{u \subset A\}}(B) = E_{\{u \subset A\}} P_u(B). \tag{1}$$

It is easy to prove that if the set $E^{(u)}$ of all possible values of u is finite, equation (1) holds true for any choice of A (when $P_u(B)$ is defined as in § 6, Chap. I). In the general case (in which $P_u(B)$ is not yet defined) we shall prove that there always exists one and only one random variable $P_u(B)$ (except for the matter of equivalence) which is defined as a function of u and which satisfies equation (1) for every choice of A from $\mathfrak{F}^{(u)}$ such that

$P^{(u)}(A) > 0$. *The function* $P_u(B)$ *of* u *thus determined to within equivalence, we call the conditional probability of* B *with respect to* u (or, for a given u). The value of $P_u(B)$ when $u = a$ we shall designate by $P_u(a; B)$.

The proof of the existence and uniqueness of $P_u(B)$. If we multiply (1) by $P\{u \subset A\} = P^{(u)}(A)$, we obtain, on the left,

$$P\{u \subset A\} P_{u \subset A}(B) = P(B\{u \subset A\}) = P(B u^{-1}(A))$$

and, on the right,

$$P\{u \subset A\} E_{\{u \subset A\}} P_u(B) = \int_{\{u \subset A\}} P_u(B) \, P(dE) = \int_A P_u(B) \, P^{(u)}(dE^{(u)}),$$

leading to the formula

$$P(B u^{-1}(A)) = \int_A P_u(B) \, P^{(u)}(dE^{(u)}); \qquad (2)$$

and conversely (1) follows from (2). In the case $P^{(u)}(A) = 0$, in which case (1) is meaningless, equation (2) becomes trivially true. Condition (2) is thus equivalent to (1). In accordance with Property IX of the integral (§ 1, Chap. IV) the random variable x is uniquely defined (except for equivalence) by means of the values of the integral

$$\int_A x \, P \, d(E)$$

for all sets of $\mathfrak{F}$. Since $P_u(B)$ is a random variable determined on the probability field $(\mathfrak{F}^{(u)}, P^{(u)})$, it follows that formula (2) uniquely determines this variable $P_u(B)$ except for equivalence.

We must still prove the existence of $P_u(B)$. We shall apply here the following theorem of Nikodym[1]:

Let $\mathfrak{F}$ be a Borel field, $P(A)$ a non-negative completely additive set function defined on $\mathfrak{F}$ (in the terminology of the probability theory, a random variable on $(\mathfrak{F}, P)$), and let $Q(A)$ be another completely additive set function defined on $\mathfrak{F}$, such that from $Q(A) \neq 0$ follows the inequality $P(A) > 0$. Then there exists a function $f(\xi)$ (in the terminology of the theory of probability, a random variable) which is measurable with respect to $\mathfrak{F}$, and which satisfies, for each set A of $\mathfrak{F}$, the equation

---

[1] O. NIKODYM, *Sur une généralisation des intégrales de M. J. Ra don*, Fund. Math. v. 15, 1930 p. 168 (Theorem III).

$$Q(A) = \int_A f(\xi)\, \mathsf{P}\,(dE).$$

In order to apply this theorem to our case, we need to prove 1° that

$$Q(A) = \mathsf{P}(Bu^{-1}(A))$$

is a completely additive function on $\mathfrak{F}^{(u)}$, 2°. that from $Q(A) \neq 0$ follows the inequality $\mathsf{P}^{(u)}(A) > 0$.

Firstly, 2° follows from

$$0 \leqq \mathsf{P}(B\,u^{-1}(A)) \leqq \mathsf{P}(u^{-1}(A)) = \mathsf{P}^{(u)}(A).$$

For the proof of 1° we set

$$A = \sum_n A_n.$$

then

$$u^{-1}(A) = \sum_n u^{-1}(A_n)$$

and

$$B\,u^{-1}(A) = \sum_n B\,u^{-1}(A_n).$$

Since P is completely additive, it follows that

$$\mathsf{P}(B\,u^{-1}(A_n)) = \sum_n \mathsf{P}(B\,u^{-1}(A_n)),$$

which was to be proved.

From the equation (1) follows an important formula (if we set $A = E^{(u)}$):

$$\mathsf{P}(B) = \mathsf{E}(\mathsf{P}_u(B)). \tag{3}$$

Now we shall prove the following two fundamental properties of conditional probability.

THEOREM I. *It is almost sure that*

$$0 \leqq \mathsf{P}_u(B) \leqq 1. \tag{4}$$

THEOREM II. *If B is decomposed into at most a countable number of sets B_n:*

$$B = \sum_n B_n,$$

then the following equality holds almost surely:

$$\mathsf{P}_u(B) = \sum_n \mathsf{P}_u(B_n). \tag{5}$$

These two properties of $\mathsf{P}_u(B)$ correspond to the two characteristic properties of the probability function $\mathsf{P}(B)$: that $0 \leqq \mathsf{P}(B) \leqq 1$ always, and that $\mathsf{P}(B)$ is completely additive. These allow us to carry over many other basic properties of the absolute probability $P(B)$ to the conditional probability $P_u(B)$. However, we must not forget that $P_u(B)$ is,for a fixed set B, a random variable determined uniquely only to within equivalence.

Proof of Theorem I. If we assume—contrary to the assertion to be proved—that on a set $M \subset E^{(u)}$ with $P^{(u)}(M) > 0$, the inequality $P_u(B) \geqq 1 + \varepsilon$, $\varepsilon > 0$, holds true, then according to formula (1)

$$P_{\{u \in M\}}(B) = E_{\{u \in M\}} P_u(B) \geqq 1 + \varepsilon,$$

which is obviously impossible. In the same way we prove that almost surely $P_u(B) \geqq 0$.

Proof of Theorem II. From the convergence of the series

$$\sum_n E \, |P_u(B_n)| = \sum_n E \, (P_u(B_n)) = \sum_n P \, (B_n) = P \, (B)$$

it follows from Property V of mathematical expectation (Chap. IV, § 2) that the series

$$\sum_n P_u(B_n)$$

almost surely converges. Since the series

$$\sum_n E_{\{u \in A\}} |P_u(B_n)| = \sum_n E_{\{u \in A\}} (P_u(B_n)) = \sum_n P_{\{u \in A\}} (B_n) = P_{\{u \in A\}}(B)$$

converges for every choice of the set A such that $P^{(u)}(A) > 0$, then from Property V of mathematical expectation just referred to it follows that for each A of the above kind we have the relation

$$E_{\{u \in A\}} \left(\sum_n P_u(B_n) \right) = \sum_n E_{\{u \in A\}} (P_u(B_n)) = P_{\{u \in A\}}(B) = E_{\{u \in A\}} (P_u(B_n)),$$

and from this, equation (5) immediately follows.

To close this section we shall point out two particular cases. If, first, $u(\xi) = c$ (a constant), then $P_c(A) = P(A)$ almost surely. If, however, we set $u(\xi) = \xi$, then we obtain at once that $P_\xi(A)$ is almost surely equal to one on A and is almost surely equal to zero on $\bar{A}$. $P_\xi(A)$ is thus revealed to be the *characteristic function* of set A.

§ 2. Explanation of a Borel Paradox

Let us choose for our basic set E the set of all points on a spherical surface. Our $\mathfrak{F}$ will be the aggregate of all Borel sets of the spherical surface. And finally, our $P(A)$ is to be proportional to the measure of set A. Let us now choose two diametrically opposite points for our poles, so that each meridian circle will be uniquely defined by the longitude $\psi, 0 \leqq \psi < \pi$. Since ψ varies from 0 only to π, — in other words, we are considering *complete* meridian circles (and not merely semicircles) — the latitude Θ must vary from $-\pi$ to $+\pi$ (and not from $-\frac{\pi}{2}$ to $+\frac{\pi}{2}$). Borel set the following problem: Required to determine "the conditional probability distribution" of latitude Θ, $-\pi \leqq \Theta < +\pi$, for a given longitude ψ.

It is easy to calculate that

$$P_\psi \{\Theta_1 \leqq \Theta < \Theta_2\} = \tfrac{1}{4} \int\limits_{\Theta_1}^{\Theta_2} |\cos\Theta|\, d\Theta .$$

The probability distribution of Θ for a given ψ is not uniform.

If we assume that the conditional probability distribution of Θ "with the hypothesis that ξ lies on the given meridian circle" must be uniform, then we have arrived at a contradiction.

This shows that the concept of a conditional probability with regard to an isolated given hypothesis whose probability equals 0 is inadmissible. For we can obtain a probability distribution for Θ on the meridian circle only if we regard this circle as an element of the decomposition of the entire spherical surface into meridian circles with the given poles.

§ 3. Conditional Probabilities with Respect to a Random Variable

If x is a random variable and $P_x(B)$ as a function of x is measurable in the Borel sense, then $P_x(B)$ can be defined in an elementary way. For we can rewrite formula (2) in § 1, to look as follows:

$$P(B)\, P_B^{(x)}(A) = \int\limits_A P_x(B)\, P^{(x)}(dE) . \qquad (1)$$

In this case we obtain from (1) at once that

$$P(B)\, F_B^{(x)}(a) = \int\limits_{-\infty}^{a} P_x(a;B)\, dF^{(x)}(a) . \qquad (2)$$

In accordance with a theorem of Lebesgue[2] it follows from (2) that

$$P_x(a;B) = P(B) \lim \frac{F_B^{(x)}(a+h) - F_B^{(x)}(a)}{F^{(x)}(a+h) - F^{(x)}(a)} \qquad h \to 0 \qquad (3)$$

which is always true except for a set H of points a for which $P^{(x)}(H) = 0$.

[2] Lebesgue, *l. c.*, 1928, pp. 301-302.

$P_x(a; B)$ was defined in § 1 except on a set G, which is such that $P^{(x)}(G) = 0$. If we now regard formula (3) as the definition of $P_x(a; B)$ (setting $P_x(a; B) = 0$ when the limit in the right hand side of (3) fails to exist), then this new variable satisfies all requirements of § 1.

If, besides, the probability densities $f^{(x)}(a)$ and $f_B^{(x)}(a)$ exist and if $f^{(x)}(a) > 0$, then formula (3) becomes

$$P_x(a; B) = P(B) \frac{f_B^{(x)}(a)}{f^{(x)}(a)}. \tag{4}$$

Moreover, from formula (3) it follows that the existence of a limit in (3) and of a probability density $f^{(x)}(a)$ results in the existence of $f_B^{(x)}(a)$. In that case

$$P(B) f_B^{(x)}(a) \leqq f^{(x)}(a). \tag{5}$$

If $P(B) > 0$, then from (4) we have

$$f_B^{(x)}(a) = \frac{P_x(a; B) f^{(x)}(a)}{P(B)}. \tag{6}$$

In case $f^{(x)}(a) = 0$, then according to (5) $f_B^{(x)}(a) = 0$ and therefore (6) also holds. If, besides, the distribution of x is continuous, we have

$$P(B) = E(P_x(B)) = \int_{-\infty}^{+\infty} P_x(a; B) \, dF^{(x)}(a) = \int_{-\infty}^{+\infty} P_x(a; B) f^{(x)}(a) \, da. \tag{7}$$

From (6) and (7) we obtain

$$f_B^{(x)}(a) = \frac{P_x(a; B) f^{(x)}(a)}{\int_{-\infty}^{+\infty} P_x(a; B) f^{(x)}(a) \, da}. \tag{8}$$

This equation gives us the so-called *Bayes' Theorem for continuous distributions*. The assumptions under which this theorem is proved are these: $P_x(B)$ is measurable in the Borel sense and at the point a is defined by formula (3), the distribution of x is continuous, and at the point a there exists a probability density $f^{(x)}(a)$.

§ 4. Conditional Mathematical Expectations

Let u be an arbitrary function of ξ, and y a random variable. The random variable $E_u(y)$, representable as a function of u and satisfying, for any set A of $\mathfrak{F}^{(u)}$ with $P^{(u)}(A) > 0$, the condition

$$\mathsf{E}_{\{u \subset A\}}(y) = \mathsf{E}_{\{u \subset A\}} \mathsf{E}_u(y) \;, \tag{1}$$

is called (if it exists) *the conditional mathematical expectation of the variable y for known value of u.*

If we multiply (1) by $P^{(u)}(A)$, we obtain

$$\int\limits_{\{u \subset A\}} y\, \mathsf{P}(dE) = \int\limits_{A} \mathsf{E}_u(y)\, \mathsf{P}^{(u)}(d\,E^{(u)}) \;. \tag{2}$$

Conversely from (2) follows formula (1). In case $\mathsf{P}^{(u)}(A) = 0$, in which case (1) is meaningless, (2) becomes trivial. In the same manner as in the case of conditional probability (§ 1) we can prove that $\mathsf{E}_u(y)$ is determined uniquely—except for equivalence—by (2).

The value of $\mathsf{E}_u(y)$ for $u = a$ we shall denote by $\mathsf{E}_u(a; y)$. Let us also note that $\mathsf{E}_u(y)$, as well as $\mathsf{P}_u(y)$, depends only upon the partition $\mathfrak{A}_u$ and may be designated by $\mathsf{E}_{\mathfrak{A}_u}(y)$.

The existence of $\mathsf{E}(y)$ is implied in the definition of $\mathsf{E}_u(y)$ (if we set $A = E^{(u)}$, then $\mathsf{E}_{\{u \subset A\}}(y) = \mathsf{E}(y)$).

We shall now prove that *the existence of* $\mathsf{E}(y)$ *is also sufficient for the existence of* $\mathsf{E}_u(y)$. For this we only need to prove that by the theorem of Nikodym (§ 1), the set function

$$Q(A) = \int\limits_{\{u \subset A\}} y\, \mathsf{P}(dE)$$

is completely additive on $\mathfrak{F}^{(u)}$ and absolutely continuous with respect to $\mathsf{P}^{(u)}(A)$. The first property is proved verbatim as in the case of conditional probability (§ 1). The second property—absolute continuity—is contained in the fact that from $Q(A) \neq 0$ the inequality $\mathsf{P}^{(u)}(A) > 0$ must follow. If we assume that $\mathsf{P}^{(u)}(A) = \mathsf{P}\{u \subset A\} = 0$, it is clear that

$$Q(A) = \int\limits_{\{u \subset A\}} y\, \mathsf{P}(dE) = 0 \;,$$

and our second requirement is thus fulfilled.

If in equation (1) we set $A = E^{(u)}$, we obtain the formula

$$\mathsf{E}(y) = \mathsf{E}\, \mathsf{E}_u(y) \;. \tag{3}$$

We can show further that almost surely

$$\mathsf{E}_u(ay + bz) = a\mathsf{E}_u(y) + b\mathsf{E}_u(z) \;, \tag{4}$$

where a and b are two arbitrary constants. (The proof is left to the reader.)

If u and v are two functions of the elementary event ξ, then the couple (u, v) can always be regarded as a function of ξ. The following important equation then holds:

$$\mathsf{E}_u \mathsf{E}_{(u,v)}(y) = \mathsf{E}_u(y). \qquad (5)$$

For, $\mathsf{E}_u(y)$ is defined by the relation

$$\mathsf{E}_{\{u \subset A\}}(y) = \mathsf{E}_{\{u \subset A\}} \mathsf{E}_u(y).$$

Therefore we must show that $\mathsf{E}_u \mathsf{E}_{(u,v)}(y)$ satisfies the equation

$$\mathsf{E}_{\{u \subset A\}}(y) = \mathsf{E}_{\{u \subset A\}} \mathsf{E}_u \mathsf{E}_{(u,v)}(y). \qquad (6)$$

From the definition of $\mathsf{E}_{(u,v)}(y)$ it follows that

$$\mathsf{E}_{\{u \subset A\}}(y) = \mathsf{E}_{\{u \subset A\}} \mathsf{E}_{(u,v)}(y). \qquad (7)$$

From the definition of $\mathsf{E}_u \mathsf{E}_{(u,v)}(y)$ it follows, moreover, that

$$\mathsf{E}_{\{u \subset A\}} \mathsf{E}_{(u,v)}(y) = \mathsf{E}_{\{u \subset A\}} \mathsf{E}_u \mathsf{E}_{(u,v)}(y). \qquad (8)$$

Equation (6) results from equations (7) and (8) and thus proves our statement.

If we set $y = \mathsf{P}_u(B)$ equal to one on B and to zero outside of B, then

$$\mathsf{E}_u(y) = \mathsf{P}_u(B),$$
$$\mathsf{E}_{(u,v)}(y) = \mathsf{P}_{(u,v)}(B).$$

In this case, from formula (5) we obtain the formula

$$\mathsf{E}_u \mathsf{P}_{(u,v)}(B) = \mathsf{P}_u(B). \qquad (9)$$

The conditional mathematical expectation $\mathsf{E}_u(y)$ may also be defined directly by means of the corresponding conditional probabilities. To do this we consider the following sums:

$$S_\lambda(u) = \sum_{k=-\infty}^{k=+\infty} k\lambda \mathsf{P}_u\{k\lambda \leqq y < (k+1)\lambda\} = \sum_k R_k. \qquad (10)$$

If $\mathsf{E}(y)$ exists, the series (10) almost certainly* converges. For we have from formula (3), of §1,

$$\mathsf{E}|R_k| = |k\lambda| \mathsf{P}\{k\lambda \leqq y < (k+1)\lambda\},$$

and the convergence of the series

$$\sum_{k=-\infty}^{k=+\infty} |k\lambda| \mathsf{P}\{k\lambda \leqq y < (k+1)\lambda\} = \sum_k \mathsf{E}|R_k|$$

---

* We use *almost certainly* interchangeably with *almost surely*.

is the necessary condition for the existence of $E(y)$ (see Chap. IV, § 1). From this convergence it follows that the series (10) converges almost certainly (see Chap. IV, § 2, V). We can further show, exactly as in the theory of the Lebesgue integral, that from the convergence of (10) for some λ, its convergence for every λ follows, and that in the case where series (10) converges, $S_\lambda(u)$ tends to a definite limit as $\lambda \to 0^3$. We can then define

$$E_u(y) = \lim_{\lambda \to 0} S_\lambda(u) \,. \tag{11}$$

To prove that the conditional expectation $E_u(y)$ defined by relation (11) satisfies the requirements set forth above, we need only convince ourselves that $E_u(y)$, as determined by (11), satisfies equation (1). We prove this fact thus:

$$E_{\{u \subset A\}} E_u(y) = \lim_{\lambda \to 0} E_{\{u \subset A\}} S_\lambda(u)$$

$$= \lim_{\lambda \to 0} \sum_{k=-\infty}^{+\infty} k\lambda P_{\{u \subset A\}}\{k\lambda \leqq y < (k+1)\lambda\} = E_{\{u \subset A\}}(y) \,.$$

The interchange of the mathematical expectation sign with the limit sign is admissible in this computation, since $S_\lambda(u)$ converges uniformly to $E_u(y)$ as $\lambda \to 0$ (a simple result of Property V of mathematical expectation in § 2). The interchange of the mathematical expectation sign and the summation sign is also admissible since the series

$$\sum_{k=-\infty}^{k=+\infty} E_{\{u \subset A\}}\{|k\lambda| P_u[k\lambda \leqq y < (k+1)\lambda]\}$$

$$= \sum_{k=-\infty}^{k=+\infty} |k\lambda| P_{\{u \subset A\}}[k\lambda \leqq y < (k+1)\lambda]$$

converges (an immediate result of Property V of mathematical expectation).

Instead of (11) we may write

$$E_u(y) = \int_E y\, P_u(dE) \,. \tag{12}$$

We must not forget here, however, that (12) is not an integral

---

[3] In this case we consider only a countable sequence of values of λ; then all probabilities $P_u\{k\lambda \leqq y < (k+1)\lambda\}$ are almost certainly defined for all these values of λ.

in the sense of § 1, Chap. IV, so that (12) is only a symbolic expression.

If x is a random variable then we call the function of x and a

$$F_x^{(y)}(a) = \mathsf{P}_x(y < a)$$

the *conditional distribution function of y for known x.*

$F_x^{(y)}(a)$ is almost certainly defined for every a. If $a < b$ then almost certainly

$$F_x^{(y)}(a) \leqq F_x^{(y)}(b).$$

From (11) and (10) it follows[4] that almost certainly

$$\mathsf{E}_x(y) = \lim_{\lambda \to 0} \sum_{k=-\infty}^{k=+\infty} k\lambda \left[F_x^{(y)}((k+1)\lambda) - F_x^{(y)}(k\lambda) \right]. \tag{13}$$

This fact can be expressed symbolically by the formula

$$\mathsf{E}_x(y) = \int_{-\infty}^{+\infty} a \, d \, F_x^{(y)}(a) \tag{14}$$

By means of the new definition of mathematical expectation [(10) and (11)] it is easy to prove that, for a real function of u,

$$\mathsf{E}_u[f(u) \, y] = f(u) \, \mathsf{E}_u(y). \tag{15}$$

---

[4] Cf. footnote 3.

Chapter VI

INDEPENDENCE; THE LAW OF LARGE NUMBERS

§ 1. Independence

DEFINITION 1 : Two functions, u and v of ξ, are *mutually independent* if for any two sets, A of $\mathfrak{F}^{(u)}$, and B of $\mathfrak{F}^{(v)}$, the following equation holds:

$$\mathsf{P}(u \subset A, v \subset B) = \mathsf{P}(u \subset A)\, \mathsf{P}(v \subset B) = \mathsf{P}^{(u)}(A)\, \mathsf{P}^{(v)}(B) . \qquad (1)$$

If the sets $E^{(u)}$ and $E^{(v)}$ consist of only a finite number of elements,

$$E^{(u)} = u_1 + u_2 + \cdots + u_n ,$$
$$E^{(v)} = v_1 + v_2 + \cdots + v_m ,$$

then our definition of independence of u and v is identical with the definition of independence of the partitions

$$E = \sum_k \{u = u_k\} ,$$
$$E = \sum_k \{v = v_k\}$$

as in § 5, Chap. I.

For the independence of u and v, the following condition is necessary and sufficient. For any choice of set A in $\mathfrak{F}^{(u)}$ the following equation holds almost certainly:

$$\mathsf{P}_v(u \subset A) = \mathsf{P}(u \subset A) . \qquad (2)$$

In the case $\mathsf{P}^{(v)}(B) = 0$, both equations (1) and (2) are satisfied, and therefore we need only prove their equivalence in the case $\mathsf{P}^{(v)}(B) > 0$. In this case (1) is equivalent to the relation

$$\mathsf{P}_{\{v \in B\}}(u \subset A) = \mathsf{P}(u \subset A) \qquad (3)$$

and therefore to the relation

$$E_{\{v \in B\}} \mathsf{P}_v(u \subset A) = \mathsf{P}(u \subset A) . \qquad (4)$$

On the other hand, it is obvious that equation (4) follows from

(2). Conversely since $P_v(u \subset A)$ is uniquely determined by (4) to within probability zero, then equation (2) follows from (4) almost certainly.

DEFINITION 2: Let M be a set of functions $u_\mu(\xi)$ of ξ. These functions are called mutually independent in their totality if the following condition is satisfied. Let M' and M'' be two non-intersecting subsets of M, and let A' (or A'') be a set from $\mathfrak{F}$ defined by a relation among u_μ from M' (or M''); then we have

$$P(A'A'') = P(A') P(A'').$$

The aggregate of all u_μ of M' (or of M'') can be regarded as coordinates of some function u' (or u''). Definition 2 requires only the independence of u' and u'' in the sense of Definition 1 for each choice of non-intersecting sets M' and M''.

If $u_1, u_2, \ldots, u_n$ are mutually independent, then in all cases

$$\left. \begin{aligned} &P\{u_1 \subset A_1, u_2 \subset A_2, \ldots, u_n \subset A_n\} \\ &= P(u_1 \subset A_1) P(u_2 \subset A_2) \ldots P(u_n \subset A_n), \end{aligned} \right\} \quad (5)$$

provided the sets A_k belong to the corresponding $\mathfrak{F}^{(u_k)}$ (proved by induction). This equation is not in general, however, at all sufficient for the mutual independence of $u_1, u_2, \ldots, u_n$.

Equation (5) is easily generalized for the case of a countably infinite product.

From the mutual independence of u_{μ_λ} in each finite group $(u_{\mu_1}, u_{\mu_2}, \ldots, u_{\mu_k})$ it does not necessarily follow that all u_μ are mutually independent.

Finally, it is easy to note that the mutual independence of the functions u_μ is in reality a property of the corresponding partitions $\mathfrak{A}_{u_\mu}$. Further, if u'_μ are single-valued functions of the corresponding u_μ, then from the mutual independence of u_μ follows that of u'_μ.

§ 2. Independent Random Variables

If $x_1, x_2, \ldots, x_n$ are mutually independent random variables then from equation (2) of the foregoing paragraph follows, in particular, the formula

$$F^{(x_1, x_2, \ldots, x_n)}(a_1, a_2, \ldots, a_n) = F^{(x_1)}(a_1) F^{(x_2)}(a_2) \ldots F^{(x_n)}(a_n). \quad (1)$$

If in this case the field $\mathfrak{F}^{(x_1, x_2, \ldots, x_n)}$ consists only of Borel sets of *the space R^n, then condition* (1) *is also sufficient for the mutual independence of the variables* $x_1, x_2, \ldots, x_n$.

Proof. Let $x' = (x_{i_1}, x_{i_2}, \ldots, x_{i_k})$ and $x'' = (x_{j_1}, x_{j_2}, \ldots, x_{j_m})$ be two non-intersecting subsystems of the variables $x_1, x_2, \ldots, x_n$. We must show, on the basis of formula (1), that for every two Borel sets A' and A'' of R^k (or R^m) the following equation holds:

$$\mathsf{P}\,(x' \subset A',\, x'' \subset A'') = \mathsf{P}\,(x' \subset A')\,\mathsf{P}\,(x'' \subset A'')\,. \qquad (2)$$

This follows at once from (1) for the sets of the form

$$A' = \{x_{i_1} < a_1,\, x_{i_2} < a_2,\, \ldots,\, x_{i_k} < a_k\},$$
$$A'' = \{x_{j_1} < b_1,\, x_{j_2} < b_2,\, \ldots,\, x_{j_m} < b_m\}.$$

It can be shown that this property of the sets A' and A'' is preserved under formation of sums and differences, from which equation (2) follows for all Borel sets.

Now let $x = \{x_\mu\}$ be an arbitrary (in general infinite) aggregate of random variables. *If the field $\mathfrak{F}^{(x)}$ coincides with the field $B\mathfrak{F}^M$ (M is the set of all μ), the aggregate of equations*

$$F_{\mu_1 \mu_2 \ldots \mu_n}(a_1, a_2, \ldots, a_n) = F_{\mu_1}(a_1)\, F_{\mu_2}(a_2) \ldots F_{\mu_n}(a_n) \qquad (3)$$

is necessary and sufficient for the mutual independence of the variables x_μ.

The necessity of this condition follows at once from formula (1). We shall now prove that it is also sufficient. Let M' and M'' be two non-intersecting subsets of the set M of all indices μ, and let A' (or A'') be a set of $B\mathfrak{F}^M$ defined by a relation among the x_μ with indices μ from M' (or M''). We must show that we then have

$$\mathsf{P}\,(A'A'') = \mathsf{P}\,(A')\,\mathsf{P}\,(A'')\,. \qquad \textbf{(4)}$$

If A' and A'' are cylinder sets then we are dealing with relations among a finite set of variables x_μ, equation (4) represents in that case a simple consequence of previous results (Formula (2)). And since relation (4) holds for sums and differences of sets A' (or A'') also, we have proved (4) for all sets of $B\mathfrak{F}^M$ as well.

Now for every μ of a set M let there be given *a priori* a distribution function $F_\mu\,(a)$; *in that case we can construct a field of probability such that certain random variables* x_μ *in that field* (μ *assuming all values in M*) *will be mutually independent, where* x_μ *will have for its distribution function the* $F_\mu\,(a)$ *given a priori.*

In order to show this it is enough to take R^M for the basic set E and $B\mathfrak{F}^M$ for the field $\mathfrak{F}$, and to define the distribution functions $F_{\mu_1\mu_2\ldots\mu_n}$ (see Chap. III, § 4) by equation (3).

Let us also note that from the mutual independence of each finite group of variables x_μ (equation (3)) there follows, as we have seen above, the mutual independence of all   on $B\mathfrak{F}^M$. In more inclusive fields of probability this property may be lost.

To conclude this section, we shall give a few more criteria for the independence of two random variables.

If two random variables x and y are mutually independent and if $E(x)$ and $E(y)$ are finite then almost certainly

$$\left.\begin{array}{l} E_x(y) = E(y), \\ E_y(x) = E(x). \end{array}\right\} \tag{5}$$

These formulas represent an immediate consequence of the second definition of conditional mathematical expectation (Formulas (10) and (11) of Chap. V, § 4). Therefore, in the case of independence both

$$f^2 = \frac{E[E(y) - E_x(y)]^2}{\sigma^2(y)} = \frac{\sigma^2[E_x(y)]}{\sigma^2(y)} \text{ and } g^2 = \frac{E[E(x) - E_y(x)]^2}{\sigma^2(x)} = \frac{\sigma^2[E_y(x)]}{\sigma^2(x)}$$

are equal to zero (provided $\sigma^2(x) > 0$ and $\sigma^2(y) > 0$). The number f^2 is called the *correlation ratio* of y with respect to x, and g^2 the same for x with respect to y (*Pearson*).

From (5) it further follows that

$$E(xy) = E(x)\,E(y) . \tag{6}$$

To prove this we apply Formula (15) of § 4, Chap. V:

$$E(xy) = E\,E_x(xy) = E[x\,E_x(y)] = E[x\,E(y)] = E(y)\,E(x) .$$

Therefore, in the case of independence

$$r = \frac{E(x,y) - E(x)\,E(y)}{\sigma(x)\,\sigma(y)}$$

is also equal to *zero*; r, as is well known, is the *correlation coefficient* of x and y.

If two random variables x and y satisfy equation (6), then they are called *uncorrelated*. For the sum

$$S = x_1 + x_2 + \ldots + x_n$$

where the $x_1, x_2, \ldots, x_n$ are uncorrelated in pairs, we can easily compute that

$$\sigma^2(s) = \sigma^2(x_1) + \sigma^2(x_2) + \cdots + \sigma^2(x_n) . \qquad (7)$$

In particular, equation (7) holds for the independent variables x_k.

§ 3. The Law of Large Numbers

Random variables s of a sequence

$$s_1, s_2, \ldots, s_n, \ldots$$

are called *stable*, if there exists a numerical sequence

$$d_1, d_2, \ldots, d_n, \ldots$$

such that for any positive ϵ

$$P\{|s_n - d_n| \geq \epsilon\}$$

converges to zero as $n \to \infty$. If all $E(s_n)$ exist and if we may set

$$d_n = E(s_n),$$

then the stability is *normal*.

If all s_n are uniformly bounded, then from

$$P\{|s_n - d_n| \geq \epsilon\} \to 0 \qquad n \to +\infty \qquad (1)$$

we obtain the relation

$$|E(s_n) - d_n| \to 0 \qquad n \to +\infty$$

and therefore

$$P\{|s_n - E(s_n)| \geq \epsilon\} \to 0. \qquad n \to +\infty \qquad (2)$$

The stability of a bounded stable sequence is thus necessarily normal.

Let

$$E(s_n - E(s_n))^2 = \sigma^2(s_n) = \sigma_n^2 .$$

According to the Tchebycheff inequality,

$$P\{|s_n - E(s_n)| \geq \epsilon\} \leq \frac{\sigma_n^2}{\epsilon^2} .$$

Therefore, the *Markov Condition*

$$\sigma_n^2 \to 0 \qquad n \to +\infty \qquad (3)$$

is sufficient for normal stability.

If $s_n - \mathsf{E}(s_n)$ are uniformly bounded:

$$| s_n - \mathsf{E}(s_n) | \leqq M,$$

then from the inequality (9) in § 3, Chap. IV,

$$\mathsf{P}\{|s_n - \mathsf{E}(s_n)| \geqq \varepsilon\} \geqq \frac{\sigma_n^2 - \varepsilon^2}{M^2}.$$

Therefore, in this case the Markov condition (3) is also necessary for the stability of the s_n.

If

$$s_n = \frac{x_1 + x_2 + \cdots + x_n}{n}$$

and the variables x_n are uncorrelated in pairs, we have

$$\sigma_n^2 = \frac{1}{n^2}\{\sigma^2(x_1) + \sigma^2(x_2) + \cdots + \sigma^2(x_n)\}.$$

Therefore, in this case, the following condition is sufficient for the normal stability of the arithmetical means s_n:

$$n^2 \sigma_n^2 = \sigma^2(x_1) + \sigma^2(x_2) + \cdots + \sigma^2(x_n) = o(n^2) \qquad (4)$$

(*Theorem of Tchebycheff*). In particular, condition (4) is ful-filled if all variables x_n are uniformly bounded.

This theorem can be generalized for the case of weakly cor-related variables x_n. If we assume that the coefficient of correla-tion r_{mn}[1] of x_m and x_n satisfies the inequality

$$r_{mn} \leqq c(|n - m|)$$

and that

$$C_n = \sum_{k=0}^{k=n-1} c(k),$$

then a sufficient condition for normal stability of the arithmetic means s is[2]

$$C_n \sigma_n^2 = o(n^2). \qquad (5)$$

In the case of *independent* summands x_n we can state a neces-sary and sufficient condition for the stability of the arithmetic means s_n. For every x_n there exists a constant m_n (the median of x_n) which satisfies the following conditions:

$$\mathsf{P}(x_n < m_n) \leqq \tfrac{1}{2},$$

$$\mathsf{P}(x_n > m_n) \leqq \tfrac{1}{2}.$$

---

[1] It is obvious that $r_{nn} = 1$ always.

[2] Cf. A. KHINTCHINE, *Sur la loi forte des grandes nombres*. C. R. de l'acad. sci. Paris v. 186, 1928, p. 285.

We set

$$x_{nk} = x_k \text{ if } | x_k - m_k | \leqq n,$$
$$x_{nk} = 0 \text{ otherwise,}$$
$$s_n^* = \frac{x_{n1} + x_{n2} + \cdots + x_{nn}}{n}.$$

Then the relations

$$\sum_{k=1}^{k=n} \mathsf{P}\{|x_k - m_k| > n\} = \sum_{k=1}^{k=n} \mathsf{P}(x_{nk} \neq x_k) \to 0, \quad n \to +\infty \quad (6)$$

$$\sigma^2(s_n^*) = \sum_{k=1}^{k=n} \sigma^2(x_{nk}) = o(n^2) \quad (7)$$

are *necessary and sufficient* for the stability of variables s_n[3].

We may here assume the constants d_n to be equal to the $\mathsf{E}(s_n^*)$ so that in the case where

$$\mathsf{E}(s_n^*) - \mathsf{E}(s_n) \to 0 \qquad\qquad n \to +\infty$$

(and only in this case) the stability is *normal*.

A further generalization of Tchebycheff's theorem is obtained if we assume that the s_n depend in some way upon the results of any n trials,

$$\mathfrak{A}_1, \mathfrak{A}_2, \ldots, \mathfrak{A}_n ,$$

so that after each definite outcome of all these n trials s_n assumes a definite value. The general idea of all these theorems known as *the law of large numbers*, consists in the fact that if the dependence of variables s_n upon each separate trial $\mathfrak{A}_k$ $(k = 1, 2, \ldots, n)$ is very small for a large n, then the variables s_n are stable. If we regard

$$\beta_{nk}^2 = \mathsf{E}[\mathsf{E}_{\mathfrak{A}_1\mathfrak{A}_2\ldots\mathfrak{A}_k}(s_n) - \mathsf{E}_{\mathfrak{A}_1\mathfrak{A}_2\ldots\mathfrak{A}_{k-1}}(s_n)]^2$$

as a reasonable measure of the dependence of variables s_n upon the trial $\mathfrak{A}_k$, then the above-mentioned general idea of the law of large numbers can be made concrete by the following considerations[4].

Let
$$z_{nk} = \mathsf{E}_{\mathfrak{A}_1\mathfrak{A}_2\ldots\mathfrak{A}_k}(s_n) - \mathsf{E}_{\mathfrak{A}_1\mathfrak{A}_2\ldots\mathfrak{A}_{k-1}}(s_n) .$$

[3] Cf. A. KOLMOGOROV. *Über die Summen durch den Zufall bestimmter unabhängiger Grössen*, Math. Ann. v. 99, 1928, pp. 309-319 (corrections and notes to this study, v. 102, 1929 pp. 484-488, Theorem VIII and a supplement on p. 318).

[4] Cf. A. KOLMOGOROV. *Sur la loi des grandes nombres.* Rend. Accad. Lincei v. 9, 1929 pp. 470-474.

Then
$$s_n - \mathsf{E}(s_n) = z_1 + z_2 + \cdots + z_n,$$

$$\mathsf{E}(z_{nk}) = \mathsf{E}\,\mathsf{E}_{\mathfrak{A}_1\mathfrak{A}_2\ldots\mathfrak{A}_k}(s_n) - \mathsf{E}\,\mathsf{E}_{\mathfrak{A}_1\mathfrak{A}_2\ldots\mathfrak{A}_{k-1}}(s_n) = \mathsf{E}(s_n) - \mathsf{E}(s_n) = 0,$$

$$\sigma^2(z_{nk}) = \mathsf{E}(z_{nk}^2) = \beta_{nk}^2.$$

We can easily compute also that the random variables z_{nk} ($k = 1, 2, \ldots, n$) are uncorrelated. For let $i < k$; then[5]

$$\mathsf{E}_{\mathfrak{A}_1\mathfrak{A}_2\ldots\mathfrak{A}_{k-1}}(z_{ni}z_{nk}) = z_{ni}\,\mathsf{E}_{\mathfrak{A}_1\mathfrak{A}_2\ldots\mathfrak{A}_{k-1}}(z_{nk})$$

$$= z_{ni}\,\mathsf{E}_{\mathfrak{A}_1\mathfrak{A}_2\ldots\mathfrak{A}_{k-1}}[\mathsf{E}_{\mathfrak{A}_1\mathfrak{A}_2\ldots\mathfrak{A}_k}(s_n) - \mathsf{E}_{\mathfrak{A}_1\mathfrak{A}_2\ldots\mathfrak{A}_{k-1}}(s_n)]$$

$$= z_{ni}[\mathsf{E}_{\mathfrak{A}_1\mathfrak{A}_2\ldots\mathfrak{A}_{k-1}}(s_n) - \mathsf{E}_{\mathfrak{A}_1\mathfrak{A}_2\ldots\mathfrak{A}_{k-1}}(s_n)] = 0$$

and therefore
$$\mathsf{E}(z_{ni}z_{nk}) = 0.$$
We thus have
$$\sigma^2(s_n) = \sigma^2(z_{n1}) + \sigma^2(z_{n2}) + \cdots + \sigma^2(z_{nn}) = \beta_{n1}^2 + \beta_{n2}^2 + \cdots + \beta_{nn}^2.$$
Therefore, the condition
$$\beta_{n1}^2 + \beta_{n2}^2 + \cdots + \beta_{nn}^2 \to 0 \qquad n \to +\infty$$
is sufficient for the normal stability of the variables s_n.

§ 4. Notes on the Concept of Mathematical Expectation

We have defined the mathematical expectation of a random variable x as

$$\mathsf{E}(x) = \int_E x\,\mathsf{P}(dE) = \int_{-\infty}^{+\infty} a\,dF^{(x)}(a),$$

where the integral on the right is understood as

$$\mathsf{E}(x) = \int_{-\infty}^{+\infty} a\,dF^{(x)}(a) = \lim\int_b^c a\,dF^{(x)}(a). \qquad \begin{matrix} b \to -\infty \\ c \to +\infty \end{matrix} \qquad (1)$$

The idea suggests itself to consider the expression

$$\mathsf{E}^*(x) = \lim\int_{-b}^{+b} a\,dF^{(x)}(a) \qquad b \to +\infty \qquad (2)$$

---

[5] Application of Formula (15) in § 4, Chap. V.

as a *generalized* mathematical expectation. We lose in this case, of course, several simple properties of mathematical expectation. For example, in this case the formula

$$E(x + y) = E(x) + E(y)$$

is not always true. In this form the generalization is hardly admissible. We may add however that, with some restrictive supplementary conditions, definition (2) becomes entirely natural and useful.

We can discuss the problem as follows. Let

$$x_1, x_2, \ldots, x_n, \ldots$$

be a sequence of mutually independent variables, having the same distribution function $F^{(x)}(a) = F^{(x_n)}(a)$, $(n = 1, 2, \ldots)$ as x. Let further

$$s_n = \frac{x_1 + x_2 + \cdots + x_n}{n}$$

We now ask whether there exists a constant $E^*(x)$ such that for every $\epsilon > 0$

$$\lim P\left(|s_n - E^*(x)| > \epsilon\right) = 0, \quad n \to +\infty . \qquad (3)$$

The answer is : *If such a constant $E^*(x)$ exists, it is expressed by Formula* (2). The necessary and sufficient condition that Formula (3) hold consists in the existence of limit (2) and the relation

$$P(|x| > n) = o\left(\frac{1}{n}\right). \qquad (4)$$

To prove this we apply the theorem that condition (4) is necessary and sufficient for the stability of the arithmetic means s_n, where, in the case of stability, we may set[6]

$$d_n = \int\limits_{-n}^{+n} a \, dF^{(x)}(a) .$$

If there exists a mathematical expectation in the former sense (Formula (1)), then condition (4) is always fulfilled[7]. Since in this case $E(x) = E^*(x)$, the condition (3) actually does define a generalization of the concept of mathematical expectation. For the *generalized mathematical expectation*, Properties I - VII

---

[6] Cf. A. KOLMOGOROV , *Bemerkungen zu meiner Arbeit*, "*Über die Summen zufälliger Grössen.*" Math. Ann. v. 102, 1929, pp. 484-488, Theorem XII.
[7] *Ibid*, Theorem XIII.

(Chap. IV, § 2) still hold; in general, however, the existence of $E^*| x |$ does not follow from the existence of $E^*(x)$.

To prove that the new concept of mathematical expectation is really more general than the previous one, it is sufficient to give the following example. Set the probability density $f^{(x)}(a)$ equal to

$$f^{(x)}(a) = \frac{C}{(| a | + 2)^2 \ln (| a | + 2)},$$

where the constant C is determined by

$$\int\limits_{-\infty}^{+\infty} f^{(x)}(a)\, da = 1.$$

It is easy to compute that in this case condition (4) is fulfilled. Formula (2) gives the value

$$E^*(x) = 0,$$

but the integral

$$\int\limits_{-\infty}^{+\infty} | a |\, dF^{(x)}(a) = \int\limits_{-\infty}^{+\infty} | a |\, f^{(x)}(a)\, da$$

diverges.

§ 5. Strong Law of Large Numbers; Convergence of Series

The random variables s_n of the sequence

$$s_1, s_2, \ldots, s_n, \ldots$$

are *strongly stable* if there exists a sequence of numbers

$$d_1, d_2, \ldots, d_n, \ldots$$

such that the random variables

$$s_n - d_n$$

almost certainly tend to zero as $n \to +\infty$. From strong stability follows, obviously, ordinary stability. If we can choose

$$d_n = E(s_n),$$

then the strong stability is *normal*.

In the Tchebycheff case,

$$s_n = \frac{x_1 + x_2 + \cdots + x_n}{n},$$

where the variables x_n are mutually independent. A sufficient[8] condition for the normal strong stability of the arithmetic means s_n is the convergence of the series

$$\sum_{n=1}^{\infty} \frac{\sigma^2(x_n)}{n^2} \,. \tag{1}$$

This condition is the best in the sense that for any series of constants b_n such that

$$\sum_{n=1}^{\infty} \frac{b_n}{n^2} = +\infty \,,$$

we can build a series of mutually independent random variables x_n such that

$$\sigma^2(x_n) = b_n$$

and the corresponding arithmetic means s_n will not be strongly stable.

If all x_n have the same distribution function $F^{(x)}(a)$, then the existence of the mathematical expectation

$$E(x) = \int_{-\infty}^{+\infty} a \, dF^{(x)}(a)$$

is necessary and sufficient for the strong stability of s_n; the stability in this case is always normal[9].

Again, let

$$x_1, x_2, \ldots, x_n, \ldots$$

be mutually independent random variables. Then the probability of convergence of the series

$$\sum_{n=1}^{\infty} x_n \tag{2}$$

is equal either to *one* or to *zero*. In particular, this probability equals one when both series

$$\sum_{n=1}^{\infty} E(x_n) \quad \text{and} \quad \sum_{n=1}^{\infty} \sigma^2(x_n)$$

converge. Let us further assume

$$y_n = x_n \text{ in case } \lceil x_n \rceil \leqq 1,$$
$$y_n = 0 \text{ in case } |x_n| > 1.$$

---

[8] Cf. A. KOLMOGOROV,· *Sur la loi forte des grandes nombres*, C. R. Acad. Sci. Paris v. 191, 1930, pp. 910-911.

[9] The proof of this statement has not yet been published.

Then in order that series (1) converge with the probability one, it is necessary and sufficient[10] that the following series converge simultaneously :

$$\sum_{n=1}^{\infty} P\{|x_n| > 1\}, \quad \sum_{n=1}^{\infty} E(y_n) \quad \text{and} \quad \sum_{n=1}^{\infty} \sigma^2(y_n)$$

---

[10] Cf. A. KHINTCHINE and A. KOLMOGOROV, *On the Convergence of Series*, Rec. Math. Soc. Moscow, v. 32, 1925, p. 668-677.

Appendix

ZERO-OR-ONE LAW IN THE THEORY
OF PROBABILITY

We have noticed several cases in which certain limiting probabilities are necessarily equal to zero or one. For example, the probability of convergence of a series of independent random variables may assume only these two values[1]. We shall prove now a general theorem including many such cases.

THEOREM: *Let* $x_1, x_2, \ldots, x_n, \ldots$ *be any random variables and let* $f(x_1, x_2, \ldots, x_n, \ldots)$ *be a Baire function[2] of the variables* $x_1, x_2, \ldots, x_n, \ldots$ *such that the conditional probability*

$$P_{x_1, x_2, \ldots, x_n}\{f(x) = 0\}$$

of the relation

$$f(x_1, x_2, \ldots, x_n, \ldots) = 0$$

remains, when the first n *variables* $x_1, x_2, \ldots, x_n$ *are known, equal to the absolute probability*

$$P\{f(x) = 0\} \tag{1}$$

for every n. *Under these conditions the probability* (1) *equals zero or one.*

In particular, the assumptions of this theorem are fulfilled if the variables x_n are mutually independent and if the value of the function $f(x)$ remains unchanged when only a finite number of variables are changed.

Proof of the Theorem: Let us denote by A the event

$$f(x) = 0.$$

We shall also investigate the field $\mathfrak{R}$ of all events which can be defined through some relations among a finite number of vari-

---

[1] Cf. Chap. VI, § 5. The same thing is true of the probability
$$P\{s_n - d_n \to 0\}$$
in the strong law of large numbers; at least, when the variables x_n are mutually independent.

[2] A Baire function is one which can be obtained by successive passages to the limit, of sequences of functions, starting with polynomials.

ables x_n. If event B belongs to $\mathfrak{R}$, then, according to the conditions of the theorem,

$$P_B(A) = P(A). \tag{2}$$

In the case $P(A) = 0$ our theorem is already true. Let now $P(A) > 0$. Then from (2) follows the formula

$$P_A(B) = \frac{P_B(A)P(B)}{P(A)} = P(B), \tag{3}$$

and therefore $P(B)$ and $P_A(B)$ are two completely additive set functions, coinciding on $\mathfrak{R}$; therefore they must remain equal to each other on every set of the Borel extension $B\mathfrak{R}$ of the field $\mathfrak{R}$. Therefore, in particular,

$$P(A) = P_A(A) = 1,$$

which proves our theorem.

Several other cases in which we can state that certain probabilities can assume only the values one and zero, were discovered by P. Lévy. See P. LÉVY, *Sur un théorème de M. Khintchine*, Bull. des Sci. Math. v. 55, 1931, pp. 145-160, Theorem II.

BIBLIOGRAPHY

BIBLIOGRAPHY

[1]. BERNSTEIN, S.: *On the axiomatic foundation of the theory of probability.* (In Russian). Mitt. Math. Ges. Charkov, 1917, Pp. 209–274.

[2]. — *Theory of probability*, 2nd edition. (In Russian). Moscow, 1927.
Government publication RSFSR.

[1]. BOREL, E.: *Les probabilités dénombrables et leurs applications arith-
métiques.* Rend. Circ. mat. Palermo Vol. 27 (1909) Pp. 247–271.

[2]. — *Principes et formules classiques*, fasc. 1 du tome I du *Traité des
probabilités* par E. BOREL et divers auteurs. Paris: Gauthier-Villars
1925.

[3]. — *Applications à l'arithmétique et à la théorie des fonctions*, fasc. 1 du
tome II du *Traité des probabilités* par E. BOREL et divers auteurs.
Paris: Gauthier-Villars 1926.

[1]. CANTELLI, F. P.: *Una teoria astratta del Calcolo delle probabilità.* Giorn.
Ist. Ital. Attuari Vol. 3 (1932) pp. 257–265.

[2]. — *Sulla legge dei grandi numeri.* Mem. Acad. Lincei Vol. 11 (1916).

[3]. — *Sulla probabilità come limite della frequenza.* Rend. Accad. Lincei
Vol. 26 (1917) Pp. 39–45.

[1]. COPELAND H : The theory of probability from the point of view of
admissible numbers. Ann. Math. Statist. Vol. 3 (1932) Pp. 143–156.

[1]. DÖRGE, K.: *Zu der von R. von Mises gegebenen Begründung der Wahrscheinlichkeitsrechnung.* Math. Z. Vol. 32 (1930) Pp. 232–258.

[1]. FRECHET, M.: *Sur la convergence en probabilité.* Metron Vol. 8 (1930)
Pp. 1–48.

[2]. — *Recherches théoriques modernes*, fasc. 3 du tome I du *Traité des
probabilités par* E. BOREL et divers auteurs. Paris: Gauthier-Villars.

[1]. KOLMOGOROV, A.: *Über die analytischen Methoden in der Wahrscheinlichkeitsrechnung.* Math. Ann. Vol. 104 (1931) Pp. 415–458.

[2]. — *The general theory of measure and the theory of probability.* (In
Russian). Sbornik trudow sektii totshnych nauk K. A., Vol. 1 (1929)
pp. 8–21.

[1]. LÉVY, P.: *Calcul des probabilités.* Paris: Gauthier-Villars.

[1]. LOMNICKI, A.: *Nouveaux fondements du calcul des probabilités.* Fundam. Math. Vol. 4 (1923) Pp. 34–71.

[1]. MISES, R. v.: *Wahrscheinlichkeitsrechnung.* Leipzig u. Wien: Fr.
Deuticke 1931.

[2]. — *Grundlagen der Wahrscheinlichkeitsrechnung.* Math. Z. Vol. 5
(1919) pp. 52–99.

[3]. — *Wahrscheinlichkeitsrechnung, Statistik und Wahrheit.* Wien: Julius
Springer 1928.

[3']. — *Probability, Statistics and Truth* (translation of above). New York:
The MacMillan Company 1939.

[1]. REICHENBACH, H.: *Axiomatik der Wahrscheinlichkeitsrechnung.* Math.
Z. Vol. 34 (1932) Pp. 568–619.

[1]. SLUTSKY, E.: *Über stochastische Asymptoten und Grenzwerte.* Metron Vol. 5 (1925) Pp. 3–89.

[2]. — *On the question of the logical foundation of the theory of probability.* (In Russian). Westnik Statistiki, Vol. 12 (1922), pp. 13–21.

[1]. STEINHAUS, H.: *Les probabilités dénombrables et leur rapport à la théorie de la mesure.* Fundam. Math. Vol. 4 (1923) Pp. 286–310.

[1]. TORNIER, E.: *Wahrscheinlichkeitsrechnung und Zahlentheorie.* J. reine angew. Math. Vol. 160 (1929) Pp. 177–198.

[2]. — *Grundlagen der Wahrscheinlichkeitsrechnung.* Acta math. Vol. 60 (1933) Pp. 239–380.

SUPPLEMENTARY

BIBLIOGRAPHY

NOTES TO SUPPLEMENTARY BIBLIOGRAPHY

The fundamental work on the measure-theoretic approach to probability theory is A. N. Kolmogorov's *Grundbegriffe der Wahrscheinlichkeitsrechnung*, of which the present work is an English translation. It is not an overstatement to say that for the past twenty-three years most of the research work in probability has been influenced by this approach, and that the axiomatic theory advanced by Kolmogorov is considered by workers in probability and statistics to be the correct one.

The publication of Kolmogorov's *Grundbegriffe* initiated a new era in the theory of probability and its methods; and the amount of research generated by the fundamental concepts due to Kolmogorov has been very great indeed. In preparing this second edition of the English translation of Kolmogorov's monograph, it seemed desirable to give a bibliography that would in some way reflect the present status and direction of research activity in the theory of probability.

In recent years many excellent books have appeared. Three of most outstanding in this group are those by Doob [12], Feller [17], and Loève [54]. Other books dealing with general probability theory, and specialized topics in probability are: [2], [3], [6], [7], [9], [19], [23], [26], [27], [28], [34], [39], [41], [42], [47], [49], [50], [67], [70], [72]. Since these books contain many references to the literature, an attempt will be made in this bibliography to list some of the research papers that have appeared in the past few years and several that are in the course of publication.

The model developed by Kolmogorov can be briefly described as follows: In every situation (that is, an experiment, observation, etc.) in which random factors enter, there is an associated probability space or triple (Ω, ξ, p), where Ω is an abstract space (the space of elementary events), ξ is a σ-algebra of subsets of Ω (the sets of events), and $p(E)$ is a measure (the probability of the event E) defined for $E \, \epsilon \, \xi$, and satisfying the condition $p(\Omega) = 1$. The Kolmogorov model has recently been discussed by

Loś [56], who considers the use of abstract algebras and σ-algebras of sets instead of algebras and σ-algebras. Kolmogorov [44] has also considered the use of metric Boolean algebras in probability.

There are many problems, especially in theoretical physics, that do not fit into the Kolmogorov theory, the reason being that these problems involve unbounded measures. Rényi [68] has developed a general axiomatic theory of probability (which contains Kolmogorov's theory as a special case) in which unbounded measures are allowed. The fundamental concept in this theory is the conditional probability of an event. Császár [10] has studied the measure-theoretic structure of the conditional probability spaces that occur in Rényi's theory.

In another direction, examples have been given by various authors which point up the fact that Kolmogorov's theory is too general. Gnedenko and Kolmogorov [27] have introduced a more restricted concept which has been termed a perfect probability space. A perfect probability space is a triple (Ω, ξ, p) such that for any real-valued ξ-measurable function g and any linear set B for which $\{\omega : g(\omega) \, \epsilon \, B\} \, \epsilon \, \xi$, there is a Borel set $D \, \epsilon \, B$ such that $P\{\omega : g(\omega) \, \epsilon \, D\} = P\{\omega : g(\omega) \, \epsilon \, B\}$. Recently, Blackwell [5] has introduced a concept that is more restricted than that of a perfect space. The concept introduced is that of a Lusin space. A Lusin space is a pair (Ω, ξ) such that (a) ξ is separable, and (b) the range of every real-valued ξ-measurable function g on Ω is an analytic set. It has been shown that if (Ω, ξ, p) is a Lusin space and p any probability measure on ξ, then (Ω, ξ, p) is a perfect probability space.

In § 6 of Chap. I, Kolmogorov gives the definition of a Markov chain. In recent years the theory of Markov chains and processes has been one of the most active areas of research in probability. An excellent introduction to this theory is given in [17]. Other references are [2], [3], [6], [12], [19], [23], [26], [34], [39], [50], [54], [67], [70], [72]. Two papers of interest are those of Harris and Robbins [29] on the ergodic theory of Markov chains, and Chung [8] on the theory of continuous parameter processes with a denumerable number of states. The paper by Chung unifies and extends the results due to Doob (cf. [12]) and Lévy [51], [52], [53].

A number of workers in probability are utilizing the theory of semi-groups [30] in the study of Markov processes and their structural properties [63]. In this approach, due primarily to Yosida [80], a one-parameter (discrete or continuous) semi-group of operators from a Banach space to itself defines the Markov process. Hille [32] and Kato [38] have used semi-group methods to integrate the Kolmogorov differential equations, and Kendall and Reuter [40] have investigated several pathological cases arising in the theory. Feller [18] and Hille [31] have studied the parabolic differential equations arising in the continuous case. Doob [13] has employed martingale theory in the semi-group approach to one-dimensional diffusion processes. Also, Hunt [33] has studied semi-groups of (probability) measures on Lie groups.

Recently several papers have appeared which are devoted to a more abstract approach to probability and consider random variables with values in a topological space which may have an algebraic structure. In [14], [21], [22], [58], [59], and [61], problems associated with Banach-space-valued random variables are considered; and in [4] similar problems are considered for Orlicz (generalized Lebesgue) spaces. Robbins [69] has considered random variables with values in any compact topological group. Segal [75] has studied the structure of probability algebras and has used this algebraic approach to extend Kolmogorov's theorem concerning the existence of real-valued random variables having any preassigned joint distribution (cf. § 4 of Chap. III). Segal [76, Chap. 3, § 13] has also considered a non-commutative probability theory.

Prohorov [66] has studied convergence properties of probability distributions defined on Banach spaces and other function spaces. These problems have been considered also by LeCam [48] and Parzen [64].

The measure-theoretic definition and basic properties of conditional probabilities and conditional expectations have been given by Kolmogorov (Chap. IV; cf. also [12] and [54]). Using an abstract approach, S. T. C. Moy [60] has considered the properties of conditional expectation as a linear transformation of the space of all extended real-valued measurable functions on a probability space into itself. In [61] she considers the conditional expectation of Banach-space-valued random variables. Nakamura and Turamuru [62] consider an expectation as a given operation of a C^*-algebra; and Umegaki [79] considers conditional expectation as a mapping of a space of measurable operators belonging to a L_1-integrable class associated with a certain W^*-algebra into itself. The work of Umegaki is concerned with the development of a non-commutative probability theory. The results of Segal [74], Dye [15], and others, in abstract integration theory are utilized in the above studies. Other papers of interest are [1], [16], [36], and [45].

The L. Schwartz theory of distributions [73] has been utilized by Gel'fand [24] in the study of generalized stochastic processes; and by Fortet [20] and Itô [35] in the study of random distributions.

Several books devoted to the study of limit theorems in probability are available: [27], [42], [47], and [49]. In addition, [12] and [54] should be consulted. Research and review papers of interest are [11], [14], [25], [37], [46], [55], [57], [65], [71], [77], and [78].

SUPPLEMENTARY BIBLIOGRAPHY

[1] ALDA, V., *On Conditional Expectations*, Czechoslovak Math. J., Vol. 5 (1955), pp. 503–505.

[2] ARLEY, N., *On the Theory of Stochastic Processes and Their Application to the Theory of Cosmic Radiation*, Copenhagen, 1943.

[3] BARTLETT, M. S., *An Introduction to Stochastic Processes*, Cambridge, 1955.

[4] BHARUCHA-REID, A. T., *On Random Elements in Orlicz Spaces*, (Abstract) Bull. Amer. Math. Soc., Vol. 62 (1956). To appear.

[5] BLACKWELL, D., *On a Class of Probability Spaces*, Proc. Third Berkeley Symposium on Math. Statistics and Probability, Vol. 2 (1956). To appear.

[6] BLANC-LAPIERRE, A., and R. FORTET, *Théorie des fonctions aléatoires*, Paris, 1953.

[7] BOCHNER, S., *Harmonic Analysis and the Theory of Probability*, Berkeley and Los Angeles, 1955.

[8] CHUNG, K. L., *Foundations of the Theory of Continuous Parameter Markov Chains*, Proc. Third Berkeley Symposium on Math. Statistics and Probabiilty, Vol. 2 (1956). To appear.

[9] CRAMÉR, H., *Mathematical Methods of Statistics*, Princeton, 1946.

[10] CSÁSZÁR, A., *Sur le structure des espace de probabilité conditionnelle*, Acta Math. Acad. Sci. Hung., Vol. 6 (1955), pp. 337–361.

[11] DERMAN, C., and H. ROBBINS, *The Strong Law of Large Numbers when the First Moment does not Exist*, Proc. Nat. Acad. Sci. U.S.A., Vol. 41 (1955), pp. 586–587.

[12] DOOB, J. L., *Stochastic Processes*, New York, 1953.

[13] DOOB, J. L., *Martingales and One-Dimensional Diffusion*, Trans. Amer. Math. Soc., Vol. 78 (1955), pp. 168–208.

[14] DOSS, S., *Sur le théorème limite central pour des variables aléatoires dans espace de Banach*, Publ. Inst. Statist. Univ. Paris, Vol. 3 (1954), pp. 143–148.

[15] DYE, H. A., *The Radon-Nikodym Theorem for Finite Rings of Operators*, Trans. Amer. Math. Soc., 72 (1952), pp. 243–280.

[16] FABIÁN, V., *A Note on the Conditional Expectations*, Czechoslovak Math. J., Vol. 4 (1954), pp. 187–191.

[17] FELLER, W., *An Introduction to Probability Theory and Its Applications*, New York, 1950.

[18] FELLER, W., *Diffusion Processes in One Dimension*, Trans. Amer. Math. Soc., Vol. 77 (1954), pp. 1–31.

[19] FORTET, R., *Calcul des probabilités*, Paris, 1950.

[20] FORTET, R., *Random Distributions with an Application to Telephone Engineering*, Proc. Third Berkeley Symposium on Math. Statistics and Probability, Vol. 2 (1956). To appear.

[21] FORTET, R., and E. MOURIER, *Résultats complémentaires sur les éléments aléatoires prenant leurs valeurs dans un espace de Banach*, Bull. Sci. Math. (2), Vol. 78 (1954), pp. 14–30.

[22] FORTET, R., and E. MOURIER, *Les fonctions aléatoires comme éléments aléatoires dans les espace de Banach*, Stud. Math., Vol. 15 (1955), pp. 62–79.

[23] FRÉCHET, M., *Recherches théoriques modernes sur le calcul des probabilités. II. Méthode des fonctions arbitraires. Théorie des événements en chaine dans d'un nombre fini d'états possibles*, Paris, 1938.

[24] GEL'FAND, I. M., *Generalized Random Processes*, Doklady Akad. Nauk SSSR (N.S.), Vol. 100 (1955), pp. 853–856. [*In Russian.*]

[25] GIHMAN, I. L., *Some Limit Theorems for Conditional Distributions*, Doklady Akad. Nauk SSSR (N.S.), Vol. 91 (1953), pp. 1003–1006. [*In Russian.*]

[26] GNEDENKO, B. V., *Course in the Theory of Probability*, Moscow-Leningrad, 1950. [*In Russian.*]

[27] GNEDENKO, B. V., and A. N. KOLMOGOROV, *Limit Distributions for Sums of Independent Random Variables*, Translated by K. L. Chung with an appendix by J. L. Doob, Cambridge, 1954.

[28] HALMOS, P. R., *Measure Theory*, New York, 1950.

[29] HARRIS, T. E., and H. ROBBINS, *Ergodic Theory of Markov Chains Admitting an Infinite Invariant Measure*, Proc. Nat. Acad. Sci. U.S.A., Vol. 39 (1953), pp. 860–864.

[30] HILLE, E., *Functional Analysis and Semi-Groups*, New York, 1948.

[31] HILLE, E., *On the Integration Problem for Fokker-Planck's Equation in the Theory of Stochastic Processes*, Onzième congrès des math. scand. (1949), pp. 185–194.

[32] HILLE, E., *On the Integration of Kolmogoroff's Differential Equations*, Proc. Nat. Acad. Sci. U.S.A., Vol. 40 (1954), pp. 20–25.

[33] HUNT, G. A., *Semi-Groups of Measures on Lie Groups*, Trans. Amer. Math. Soc., Vol. 81 (1956), pp. 264–293.

[34] ITÔ, K., *Theory of Probability*, Tokyo, 1953.

[35] ITÔ, K., *Stationary Random Distributions*, Mem. Coll. Sci. Univ. Kyoto, Ser. A. Math., Vol. 28 (1954), pp. 209–223.

[36] JIŘINA, M., *Conditional Probabilities on Strictly Separable σ-Algebras*, Czechoslovak Math. J., Vol. 4 (1954), pp. 372–380. [*In Czech.*]

[37] KALLIANPUR, G., *On a Limit Theorem for Dependent Random Variables*, Doklady Akad. Nauk SSSR (N.S.) Vol. 101 (1955), pp. 13–16. [*In Russian.*]

[38] KATO, T., *On the Semi-Groups Generated by Kolmogoroff's Differential Equations*, J. Math. Soc. Japan, Vol. 6 (1954), pp. 1–15.

[39] KAWADA, Y., *The Theory of Probability*, Tokyo, 1952.

[40] KENDALL, D. G., and G. E. H. REUTER, *Some Pathological Markov Processes with a Denumerable Infinity of States and the Associated Semigroups of Transformations in l*, Proc. Symp. on Stochastic Processes (Amsterdam), 1954. To appear.

[41] KHINTCHINE, A., *Asymptotische Gesetze der Wahrscheinlichkeitsrechnung*, Berlin, 1933. [*Reprint*, CHELSEA PUBLISHING COMPANY.]

[42] KHINTCHINE, A., *Limit Laws of Sums of Independent Random Variables*, Moscow-Leningrad, 1938. [*In Russian.*]

[43] KOLMOGOROV, A., *Grundbegriffe der Wahrscheinlichkeitsrechnung*, Berlin, 1933. [The present work is an English translation of this.]

[44] KOLMOGOROV, A., *Algèbres de Boole métriques complètes*, VI Zjazd Mat. Pols., Warsaw (1948), pp. 21–30.

[45] KOLMOGOROV, A., *A Theorem on the Convergence of Conditional Mathematical Expectations and Some of Its Applications*, Comptes Rendus du Premier Congrès des Mathématiciens Hongrois (1952), pp. 367–386. [*In Russian and Hungarian.*]

[46] KOLMOGOROV, A., *Some Work of Recent Years in the Field of Limit Theorems in the Theory of Probability*, Vestnik Moskov. Univ. Ser. Fiz.-Mat. Estest. Nauk, Vol. 8 (1953), pp. 29–38.

[47] KUNISAWA, K., *Limit Theorems in Probability Theory*, Tokyo, 1949.

[48] LECAM, L., *Convergence in Distribution of Stochastic Processes*, Univ. California Publ. Statistics. To appear.

[49] LÉVY, P., *Théorie de l'addition des variables aléatoires*, Paris, 1937.

[50] LÉVY, P., *Processus stochastiques et mouvement Brownien*, Paris, 1948.

[51] LÉVY, P., *Systèmes markoviens et stationnaires. Cas dénombrable*, Ann. Sci. École Norm. Sup., Vol. 68 (1951), pp. 327–401.

[52] LÉVY, P., *Complément à l'étude des processus de Markoff*, Ann. Sci. École Norm. Sup., Vol. 69 (1952), pp. 203–212.

[53] LÉVY, P., *Processus markoviens et stationnaires du cinquieme type (infinité denombrable d'états possibles, paramètre continu)*, C. R. Acad. Sci. Paris, Vol. 236 (1953), pp. 1630–1632.

[54] LOÈVE, M., *Probability Theory*, New York, 1955.

[55] LOÈVE, M., *Variational Terms and the Central Limit Problem*, Proc. Third Berkeley Symposium on Math. Statistics and Probability, Vol. 2 (1956). To appear.

[56] ŁOŚ, J., *On the Axiomatic Treatment of Probability*, Colloq Math., Vol. 3 (1955), pp. 125–137.

[57] MARSAGLIA, G., *Iterated Limits and the Central Limit Theorem for Dependent Variables*, Proc. Amer. Math. Soc., Vol. 5 (1954), pp. 987–991.

[58] MOURIER, E., *Eléments aléatoires dans un espace de Banach*, Ann. Inst. H. Poincare, Vol. 13 (1953), pp. 161–244.

[59] MOURIER, E., *L-Random Elements and L-Random Elements in Banach Spaces*, Proc. Berkeley Symposium on Math. Statistics and Probability, Vol. 2 (1956). To appear.

[60] MOY, S. T. C., *Characterizations of Conditional Expectation as a Transformation on Function Spaces*, Pacific J. Math., Vol. 4 (1954), pp. 47–63.

[61] MOY, S. T. C., *Conditional Expectations of Banach Space Valued Random Variables and Their Properties*, (Abstract) Bull. Amer. Math. Soc., Vol. 62 (1956). To appear.

[62] NAKAMURA, M. and T. TURUMARU, *Expectation in an Operator Algebra*, Tôhoku Math. J., Vol. 6 (1954), pp. 182–188.

[63] NEVEU, J., *Etude des semi-groups de Markoff*, (Thesis) Paris, 1955.

[64] PARZEN, E., *Convergence in Distribution and Fourier-Stieltjes Transforms of Random Functions*,(Abstract) Ann. Math. Statistics, Vol. 26 (1955), p. 771.

[65] PARZEN, E., *A Central Limit Theorem for Multilinear Stochastic Processes*, (Abstract) Ann. Math. Statistics, Vol. 27 (1956), p. 206.

[66] PROHOROV, YU. V., *Probability Distributions in Functional Spaces*, Uspehi Matem. Nauk (N.S.), Vol. 8 (1953), pp. 165–167. [*In Russian.*]

[67] RÉNYI, A., *The Calculus of Probabilities*, Budapest, 1954. [*In Hungarian.*]

[68] RÉNYI, A., *On a New Axiomatic Theory of Probability*, Acta Math. Acad. Sci. Hung., Vol. 6 (1955), pp. 285–335.

[69] ROBBINS, H., *On the Equidistribution of Sums of Independent Random Variables*, Proc. Amer. Math. Soc., Vol. 4 (1953), pp. 786–799.

[70] ROMANOVSKI, V. I., *Discrete Markov Chains*, Moscow-Leningrad, 1949. [*In Russian.*]

[71] ROSENBLATT, M., *A Central Limit Theorem and a Strong Mixing Condition*, Proc. Nat. Acad. Sci. U.S.A., Vol. 42 (1956), pp. 43–47.

[72] SARYMSAKOV, T. A., *Elements of the Theory of Markov Processes*, Moscow, 1954. [*In Russian.*]

[73] SCHWARTZ, L., *Théorie des distributions*, Paris, 1950-51.

[74] SEGAL, I. E., *A Non-Commutative Extension of Abstract Integration*, Ann. Math. Vol. 57 (1953), pp. 401–457.

[75] SEGAL, I. E., *Abstract Probability Spaces and a Theorem of Kolmogoroff*, Amer. J. Math., Vol. 76 (1954), pp. 177–181.

[76] SEGAL, I. E., *A Mathematical Approach to Elementary Particles and Their Fields*, University of Chicago, 1955. [Mimeographed Lecture Notes.]

[77] TAKANO, K., *On Some Limit Theorems of Probability Distributions*, Ann. Inst. Statist. Math., Tokyo, Vol. 6 (1954), pp. 37–113.

[78] TSURUMI, S., *On the Strong Law of Large Numbers*, Tôhoku Math. J., Vol. 7 (1955), pp. 166–170.

[79] UMEGAKI, H., *Conditional Expectation in an Operator Algebra*, Tôhoku Math. J., Vol. 6 (1954), pp. 177–181.

[80] YOSIDA, K., *Operator Theoretical Treatment of the Markoff's Process*, Proc. Imp. Acad. Japan, Vol. 14 (1938), pp. 363–367.